TO BEE: A BUDDING BEEKEEPER'S BOOK

Beginner to Beekeeping? Discover How to Start Your First Hive in Your Own Backyard. Harvest Your Own Honey and Have Fun Doing It!

BOB JAKES

SPECIAL BONUS!

Want This Bonus Book for free?

Get FREE, unlimited access to it and all of my new books by joining the Fan base!

SCAN W/ YOUR CAMERA TO JOIN!

Contents

Introduction

Honey bees, as well as honey and other beekeeping products, have been of value and interest to human beings from time immemorial. The beneficial effects of honey-related products on the human body are worthwhile in themselves. And the strictly regulated lifestyle of bees, their tiny, alien societies, is a subject of fascination. Anyone who cares about the health of their loved ones would want to learn more about honey, royal jelly, pollen, and propolis, all of which are biologically active products of beekeeping.

Of course, any apiary has a peculiar aura: even the air filled with the buzzing of bees has a beneficial effect on the human body. But at the same time, each apiary is not the same. Only a competent beekeeper who loves the world of bees and each bee individually can create a highly profitable apiary that brings him joy and profit. This book will help you independently equip just such an apiary. The first chapters of the book are devoted to this; they outline the best methods of building up the strength of bee colonies.

The issues of health preservation and the practical application of beekeeping products useful to humans are worth studying. You will find a discussion of these subjects in the later chapters of the book. We'll look at proven methods of domestic and foreign apitherapy, folk recipes for the treatment and prevention of many common diseases with the help of beekeeping products, and instructions for the preparation of honey wines and the use of cosmetic masks and creams using honey. These sections are designed for the people interested in preserving their inner and outer beauty, as well as improving the health of their loved ones.

A chapter containing the most necessary information for novice beekeepers is included in the book. It outlines the techniques and methods of performing apiary work, proven in practice, both in simple, everyday conditions, and in those that can be called non-standard (in particular, methods of stopping bee theft, preventing swarming, etc.). Also, that chapter deals with methods and techniques for working with the most complex devices and tools. It will allow even novice beekeepers to cope with most issues, even if they lack prior experience.

In addition, the book discusses the design features of devices and tools used in beekeeping.

Blooming sunflower, flying bee, and a curious little girl

My name is Bob Jakes, and I am a keen hobbyist in the field of apiology. I have spent many years studying and experimenting with honey bees. Nowadays, I enjoy helping others set up their own hives and experiencing the benefits and joys of raising bees. Have you ever wished you could find a simple but comprehensive guide for how to get started with beekeeping? Well, you have come to the right place. And, from an economic standpoint, beekeeping can also be a great home-based business to add income to your finances.

In this book, you will find the answer to any question on practical beekeeping. And the more important the question, the more detailed the answer to it is. The book will help the beekeeper in a number of areas. You'll be able to keep bee colonies in winter, build up their strength in spring and early summer, increase the apiary, and harvest honey. The info I present will also help you maintain good relations with your neighbors - there are a lot of scare stories about bees. And the sheer pleasure you'll get from tending to a bee colony shouldn't be underestimated, either.

My main goal is to help a beginner reach a level of expertise as a beekeeper with the least amount of mistakes on his way. I want you to gain confidence and skills in looking after bees, to become, in the end, a master of your craft. In addition, I hope that my reader will become more careful with nature, will love bees, and will be able to put into practice at least some of the information in the book. And, obviously, I'd like you to enjoy real, floral (and not fake) honey.

I hope my book will help amateur beekeepers increase the profitability of apiaries, and gain confidence in their own actions. Beekeeping is a wonderful and rewarding business. The life of bees is so interesting and entertaining that a person who works in an apiary probably won't notice how quickly the day goes by - bees really are that intriguing! Involve your whole family in an interesting, environmentally friendly, and useful business. The household will thank you for this interesting and useful hobby!

ONE

Bee Biology

Meet the bees - Composition of the bee family

A bee colony is a complex organism consisting of several thousand worker bees, several hundred drones, and a queen linked into a single whole. The community is rigorously disciplined and organized. It can collect and make a lot of honey and pollen, defend itself from enemies, maintain an optimal temperature and humidity in the hive, and multiply. Each bee colony has its individual characteristics - a specific smell, degree of aggressiveness, ability to collect honey and propolize nests. Its character often depends on the nature of the queen. When an old queen is replaced with a new one, the properties of the bee colony also change: the previous generation is replaced by a new one with other hereditary properties.

The unity of the bee family is supported by complex relationships between its members. These include trophic and tactile contacts (exchange of food and pheromones), signal sounds and movements, etc. The bee colony operates as a unit and in its entirety. Each bee performs a specific function aimed at prolonging the life of the entire family.

The Queen Bee (marked "39")

A queen is an individual in a bee family capable of reproducing offspring. In size and weight, it surpasses all other bees. The length of her body depends on the breed and the season of the year and ranges from 20 to 25 mm. The mass of the fertile queen is from 200 to 250 mg, and an infertile one is from 150 to 200 mg. A fully fertile queen lays from 1000 to 2000 eggs per day, 150-200 thousand eggs per season. The queen spends 40–46 seconds laying one egg. Her weight - depending on the age of the queen, the number of bees in the family, and the season - ranges from 0.128 to 0.221 mg. Young queens lay more eggs than old ones.

Under the same conditions of keeping, the mass of eggs is in direct proportion to their number laid by the queen per day. In June (the height of oviposition), the egg weight is 0.133 mg; in July - 0.141 mg; in August - 0.163 mg.

Usually the queen lives in a family for 3-5 years. Under unfavorable wintering conditions (lack of food reserves, etc.), she dies later than the bulk of the bees. The survival rate of large queens is higher than that of small ones.

A young queen leaves for mating 7-10 days after leaving the mother's cells. By this time, her scent glands, located under the second, third, and fourth tergites of the abdomen, begin to function. Substances with a specific odor are formed. This helps attract drones during mating flights. The optimum temperature for mating is + 27-28 ° C.

The queen begins to lay eggs in February and ends in the fall, with the onset of cold weather. She lays the largest number of them in the first two years of life. With age, egg-laying decreases, and an old queen, along with fertilized eggs, will lay many unfertilized ones.

After mating, the queen becomes fertile. After 3-4 days, less often after 7 days, she begins to lay eggs. She lays eggs of two types: fertilized, in the opening of which sperm have entered, and unfertilized - there is no sperm there. If, for some reason, the queen does not mate with the drones in the first 2 weeks, she loses the ability to mate and becomes infertile. A family with such a queen will die if the beekeeper does not provide her with timely assistance.

The worker bees take good care of the queen; they clean and feed her. During feeding, the bees transfer to the queen about 66% of the food contained in their honey goiter.

Worker bees are females of the bee family with underdeveloped genitals. The body length of a bee is 12-14 mm, and the weight of a one-day bee in different breeds ranges from 90 to 115 mg. In one kg of bees, there are 10-11 thousand individuals. Their number in the family varies depending on the season: in the spring, in a strong family, there are up to 20 thousand bees; in the summer - 60–80; in the fall - up to 30 thousand.

The body temperature of bees largely depends on the outside temperature, but within certain limits, they can regulate it. Heat is generated by muscle activity. The body temperature of a bee during flight depends on the ambient temperature: at + 22–26 ° C it reaches + 35–37 ° C. At + 35–37 ° C it is +42 ° C. A bee that has completed a flight has a temperature of 6–20 ° C higher than the ambient temperature. The body cools down as a result of a decrease in metabolism, a decrease in oxygen consumption, and due to the evaporation of water.

Worker bees feed the larvae, collect nectar and pollen, build honeycombs, guard the nest, regulate the temperature and humidity in the nest, keep the hive clean, and look after the queen. The most aggressive bees are aged 7-20 days.

Long-lived bees appear in the fall, that is, during the period when there is no brood in the nests. At this time, young bees intensively feed on bee bread, which, along with a decrease or lack of work on brood feeding, contributes to the accumulation of reserve nutrients in their bodies. The live weight of bees in autumn increases compared with the summer by 13–19%, and the dry weight - by 16–26%.

Laying worker bees are worker bees that can lay unfertilized eggs. They appear in families that live without queens for a long time, as well as during swarming. The laying worker bee can lay 19 to 30 eggs. She doesn't glue eggs to the bottom of the cell, but to its walls. On this basis, it is easy to distinguish the presence of laying worker bees in the hive. Laying worker bees, in whose ovaries eggs began to develop, are called anatomical laying worker bees, and those that lay eggs are called physiological laying worker bees. The number of anatomical laying worker bees can reach 90%, and physiological - up to 25%.

The lifetime of bees is not the same and depends on the time of exit from the cell and the work performed. In a typical colony, with a queen bee hatched in March, they can live up to 35 days;

in June - up to 30; hatched during the main honey harvest - 28-30; in August and September - up to 80-100 days. In colonies without broods, bees can live up to a year.

Drones are males designed to mate with young queens. The drone's body length is 15–17 mm, and its weight is 200–250 mg. They appear in the family in May - June. Drones become sexually mature on the 8-14th day after leaving the cell. During mating with queens, drones gather in places where air currents converge, raising them 10 m from the ground. The collection points are located at a distance of 200 m to 7 km from the apiary.

During the active period, worker bees take care of the drones and feed them. Approximately 18% of drones replenish their food needs from bees by 10-30%, 62% - by 35-75%, and 10% feed from honey cells. On average, 47% of males are provided with food in the process of trophic contacts with bees. By the end of summer, the bees stop feeding the drone brood and prevent the drones from eating. Drones weakened from hunger are thrown out of the hive. Such expulsion indicates the end of the honey collection. Drones hibernate only in queenless families, or in families with infertile queens.

Bee feed

Bees collect nectar and pollen from plants. This is processed into fodder - honey and bee bread. Bees' food contains all vital nutrients - proteins, fats, carbohydrates, minerals, vitamins. Along with nectar and pollen, bees also bring water to the hive.

Bee bread

Nectar is a carbohydrate feed consisting of sugars (sucrose, fructose, and glucose), water, and a small amount of proteins and mineral salts. The content of water and sugars in nectar varies considerably depending on the floristic composition and meteorological conditions.

Bees collect nectar more actively when it contains about 50% of sugars (if it has a sugar concentration of less than 4.25%, they do not collect it at all). At the same time, the high content of sugars in nectar makes it difficult for bees to collect it. By adding enzymes of the salivary glands to the nectar and removing water from it, the insects process it into honey. In hives with normal ventilation, the evaporation of water from the nectar lasts up to 5 days. A decrease in ventilation delays the removal of water from it by more than 20 days. Removing 450 g of water requires 100 g of sugars.

As a result of the processing of nectar and honeydew by bees, honey is obtained - a mixture of fruit and grape sugar. Honey contains water (up to 17-21%), organic acids, nitrogenous and mineral substances, and vitamins. In total, there are about 300 components in it.

One bee family needs 70–90 kg of honey to maintain vital activity during a year, of which 10–12 kg is consumed in winter. During the year, the bee colony excretes about 7 kg of indigestible honey residues with feces.

Pollen is small grains that ripen in the anthers of a flower and serves as a protein feed necessary for feeding the larvae. When collecting pollen from plants, bees add saliva and nectar to it and form lumps (pollen) from it; this is placed on their hind legs in special recesses (baskets). The weight of the load is from 8 to 22 mg.

Under favorable conditions, a bee can collect pollen in 30 minutes. To collect one portion of pollen, a bee visits up to 120 flowers in the process of each flight. In the absence of nectar from pollen and/or a decrease in the numbers of the hive, bees make honey that is more 'lightweight.'

For the development of one bee, you need about 0.1 g of bee bread. In the spring and summer, growing a strong colony (60–80 thousand bees) requires 6–8 kg of bee bread. During the year, the family spends 16-18 kg of bee bread. In the summer, a strong family accumulates 15 kg. With a shortage of bee bread or pollen, bees reduce the number of larvae fed and have reduced yield.

Water is an indispensable part of the bee's body and plays an important role in metabolic processes, as well as in regulating the moisture in the nest. If nectar enters the hive, the bees usually do not need water. In spring, a family consumes an average of 100-200 ml of water per day. In hot, dry weather, the average moves up to more than 400 g.

TWO

How to Get Started

A novice beekeeper sometimes resembles a child who has directed all their energy to solving a single mystery - what is inside the toy and how does its mechanism work? He begins by checking his apiary frequently. He wants to be a conscientious and competent beekeeper. He will open the hive, get out a couple of frames, look at them carefully, and put them back in. But, let's say he didn't prepare properly and didn't put on protective clothing; it's a novice mistake that can happen.

However - the mistake has consequences. While he was turning the frame over, he was a little careless. Perhaps one of the bees stung him, and he wanted to get the inspection over quickly. He may have crushed some bees in his haste. He returns home, but doesn't realize that a serious problem has occurred. He accidentally crushed the queen.

In a few more days, he will prepare himself better and return to the hive, this time with a smoke-cutter and proper clothing. He performs a routine check again and is satisfied. But he doesn't notice the queen has died. By fall, there is no honey in the hive. Not for himself or his neighbors, not even for the bees. And the colony is facing destruction.

Bear in mind, our hypothetical keeper doesn't know he's the guilty party. So, he could blame the bees. Or he thinks it's the bee seller - perhaps there was already something wrong with the colony when he got it? A year or two of work and the keeper's attempt has ended in failure. He'll tell his neighbors that he doesn't have time for bees anymore. And he reproaches himself: "Why didn't I just buy myself a three-liter jar of honey? How much do I need?"

And he will not understand the main thing - that if he had given the bees their space at the very beginning, had not disturbed the thermal regime in the nest, hadn't distracted them from work, then he and the colony would have been fine. A bee colony lives by its own laws. A keeper shouldn't interfere or intervene if it's not necessary.

Well, experience does not come immediately. It may take several years to acquire expertise, perhaps five to fifteen years. Everyone has different abilities and everyone reaches the pinnacle of mastery in different ways.

Where to place your apiary

In beekeeping literature, there is a lot of theoretical advice on this question, but in the real world, there is only one answer. You will need to place your apiary where your house is located, where your land plot is located, and where there is free space on it.

A properly located apiary

Before deciding on an apiary, make sure that this is not going to test the patience of your neighbors. You want to keep friendly relations with them, and people can be nervous about bees. Reassuring them that bees are safe, that you know what you're doing, and that you're going to give them a jar or two of delicious honey all helps to get your neighbors on board with your new hobby. Neither you nor your neighbors want bees to cause stress. If you handle it clumsily, believe me, people who live near your home will become unhappy and annoyed. You're better off breeding fish in a pond. At least they won't be seen as dangerous, and it will be calmer and more useful for you than getting constant reproaches and accusations - or worse - in your mailbox.

The main thing is that on the apiary in the July sun, there should be a small shadow from the trees to protect the colony from overheating, although if it is not there, you can use improvised means, some hay and branches, to shelter the hives from the sun.

A two-meter fence might be an idea to enclose the apiary and make it less likely to interfere with your neighbors' normal life.

There are several objective situations when it is better not to have bees. First, if you live in the middle of a small island and there is water around you. Your bees may be in danger of drowning. If there are plenty of rival apiaries in the neighborhood, that would be a problem; how much honey will your bees collect with such competition? If there is a sugar processing plant nearby, the bees will be attracted to it and bring it back. You don't want artificial sugar finding its way into your colony. If the local fields are treated with pesticides, that's not good for the health of your bees.

And there are a few other places where you can't put hives next to. These are roads, cattle yards, kindergartens, and schools.

Give the bees some privacy and let them be a safe distance from your home. You don't want to walk around your living room with a beekeeper's dressing gown and a protective net on your head. If the entrances to the bees' colony are near your windows, that's not great. The apiary shouldn't interfere with your home, but it needs to be where you can access it quickly.

Location of hives

The hives are located at a distance of 3-4 m from each other. If they are arranged in rows, then the distance between the rows should be at least four meters. With a denser arrangement, it is necessary to paint the hives in different colors so that the bees do not confuse their hives and go to the wrong one. Beehives should be placed with entrances to the east or southeast. They should have a slight forward tilt so that rainwater does not flow inside. Ahead, on the take-off line, there should be no fence, trees, or hives.

Beehives are most often placed on pegs driven into the ground. Their height should not be higher than 25–35 cm. Boxes filled with straw or dry leaves are also used. Even rubber tires from cars are good. I use logs that are freshly cut and peeled from the bark. I bury their lower parts (about a third of the length) in the ground. The stability of such a structure has been tested by time, strong winds, and bad weather. I've also used ordinary plastic bottle crates as coasters. They do the job quite well.

Some beekeepers arrange hives in the attics of residential buildings, but this has the potential for unnecessary problems. Imagine what might happen when the whole space is covered with smoke and thousands of angry bees are hovering around. Some people might like to take the chance, but we will act as our grandfathers and great-grandfathers did - be smart and keep bees outside.

Which hive to choose

If you have no experience, if there is no benevolent teacher nearby to explain to you the methods of working with multi-body or double-body hives, I advise you to try a single-body 12-frame hive on a 435 × 300 mm frame (nesting). It is used with two store extensions in which you can put 12 semi-frames measuring 435 × 145 mm. All honey reserves located in the nest belong only to the bees and are intended for their wintering.

If necessary, it is possible to place not half-frames, but 12 full frames in the extensions, placed on top of each other. There are those who like to complicate the design and put a second body instead of two extensions, but then only full frames are placed in it. In such cases, it is used as a two-body hive that can hold 24 full frames. If necessary, an extension, one or two, is even placed between the bodies or on the upper body. After a couple of years, you can buy a multi-hive set-up. If you wish, the rest of the apiary can be expanded with their help. The disadvantages of multi-body and double-body hives include the considerable weight of the upper body. If the upper body is filled with honey, it might weigh between 30 and 60 kg. That isn't the easiest thing to lift. So, that's a factor to bear in mind.

Larger hives, especially those designed for frames 20, 22, and 24, tend to be massive, heavy, and difficult to transport to the winter house. It is good to keep them in the apiary all year round, but this option is unacceptable for many beekeepers.

You may wonder if you can come up with an original device yourself? Back in the late 1980s, there were more than 500 different hive systems. If you have a brilliant idea, you can try it out, but please get some expert advice on it first. Frankly, it is hardly worth starting beekeeping practice with a beehive of your own design. Experience shows that over time, due to many shortcomings, such a homemade product usually has to be replaced with a standard hive. There is something to be said for tradition.

Whichever type of hive you choose, remember, there should be no cracks in its structure, it should keep warm in winter, and prevent the colony from overheating in summer. Its parts must be interchangeable with parts of other hives. It should be well-colored with light colors: white, yellow, beige, gray, light green, blue (but not red). The paints themselves must be prepared with natural drying oil. You do not need to paint the hives on the inside. You will still clean it from time to time, scrape it, and even process it with a blowtorch.

THREE

Main Inventory

Get yourself a separate storage area for your tools and basic equipment. Let it be a small pantry, where only you will have the right to enter. Why is that? It's very simple; when all the objects are once arranged in order, it is easier to maintain this order than to recreate it over and over again. When someone else comes in and rearranges your tools, then you can guarantee that when you need something, it won't be where it should be. So, keep this area private and tidy.

There are not that many items of beekeeping equipment you'll need, but it is worth purchasing them before you get started.

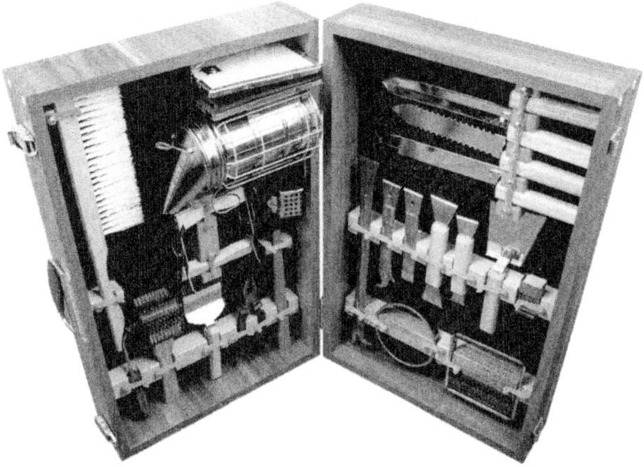

Beekeepers' toolkit

Honey extractor

You certainly can't do without this item. Without it, you will only have honeycomb honey in stock. If you want to, save on something else, but not the honey extractor. This expensive machine must be as good and accurate as possible. Take your time and get the best one you can. Ask any beekeepers you know what's best to go for. Check with shop assistants which honey extractors are in greatest demand. Try twisting a few samples with your own hands and see if this is what works for you. Otherwise, you will suffer for many years and scold yourself for being hasty. After completing the work, do not forget to rinse and dry the honey extractor thoroughly. During storage, cover it well with a cloth or oilcloth.

Box for carrying frames

The box must have a lid. The best material for making it is plywood. Size - under the standard frame. How many frames should fit in it is something to decide for yourself. Better from 6 to 8, so that it is not too heavy to carry around the site. On the side of one of the ends of the box, make a notch. Such a box can be used for carrying frames, and for collecting swarms from your apiary. It can also be useful for carrying wild bees. In winter, such a box is good for storing honey.

Smoker

Without this, it is difficult to imagine the normal work of a beekeeper. The smoke in the hive forces the bees to collect food in reserve in their honey goiter, which is why they cannot bend their abdomen to sting. While this is a reliable way to avoid a serious bee attack on the beekeeper, it is nevertheless not a good idea to overdo it. Excessive and continuous fumigation can only upset the bees. Also, after this procedure, they take a while to regain their senses and so the routine of the colony is disrupted. You'll need a smoker occasionally - just don't go crazy with it.

For a smoker, it is best to use decaying deciduous trees - aspen, alder, and poplar. Birch and conifers, rags and wool are not suitable. Their smoke only irritates the bees. Before working in the apiary, the smoker must be ignited and filled with rot, which will smolder. They need exactly such an amount so that the smoker does not go out at the most crucial moment. If this happens, you must immediately stop work, cover the hive with a canvas and a pillow and re-ignite the smoker. And only after that, resume the planned activities.

It is better not to have hot smoke in the hive, but to keep the nozzle at some distance (10-15 cm) from the bees.

Beekeeping knife, roller, or special fork

These items are used only for unpacking honeycombs. The knife should be well-sharpened and heated so as not to break the honeycomb.

Feeders

Feeders are used to induce feeding for bees when they need it. This can be in March - April, or in the last ten days of August. You need to have feeders according to the number of hives.

Spur roller

This is a cogwheel with a circular groove for guiding along the wire. It's required for rolling the wire stretched over the frame to the sheets of foundation. When working, the spur must be warmed up in a glass of hot water.

Mouse guard

Used to prevent mice from entering the hive through the entrance. You don't want creatures like these breaking into your hive.

Dividing grid

This is employed to block the passage of the queen from the nest to the extensions. It can also be used when combining several swarms into one colony in order to separate excess queens.

Beekeeping chisel

A chisel is necessary when inspecting hives and is used to peel off and move frames when cleaning the walls of the hive, etc. It's an irreplaceable item, and it is better to have one more chisel in reserve.

Hole punch or awl

It is necessary for piercing the frames of the holes in the strips through which the wire will be passed. I often use a sheet metal template. Through its holes, I make drill holes with a drill. Four on each side of the full frame and two on each side of the half frame.

Strainers for filtering honey

They are made of tinned steel; their mesh sizes are from 1 to 3 mm. The coarse mesh sieve is placed on the fine mesh sieve. They are used only for filtering and cleaning honey from wax, propolis, and other impurities.

Face mesh

Keep several of these. They should be provided to the beekeeper himself and all members of his family. Sometimes it happens that bees, either because of bad weather or because they're upset, can be aggressive. They are sensitive creatures and naturally defensive about their homes. A face mesh is a good way to keep you safe.

Protective overalls or suit

When working with bees, a protective suit or overalls is absolutely essential. Any attempt to do without them will end up with you being hurt. They should be white and clean so as not to irritate the bees with either color or smell. Don't buy suits made of fine fabric; if you do, you will have to put at least a few more thick clothes under the overalls. It is unlikely that in hot weather that this will be comfortable for you.

If you don't find a decent suit on sale, it's possible to get someone reliable to sew one for you. The cut only needs to be quite simple - the main thing is that the fabric is cotton and as dense as possible. If the suit is not ideal, bees will occasionally find a "weak link" in it and you'll end up being stung. Rubber gloves are also a necessity.

Beekeeping overalls, gloves, and frames

Buying hives

When you're starting, you don't want to get too carried away. The best number of bee colonies is two or three. Budget for this and also for the tools you'll need, the most expensive part of which will be a honey extractor.

Try in advance, preferably in the fall, to agree on the purchase of populated hives (with bees). Speak only with people who specialize in such sales - go to reputable shops and individuals. Get receipts and know that the seller is someone you can return to if there's anything wrong with the colony. Don't try to get a cheap apiary from a questionable source. If it's an older person who no longer wishes to look after bees, that's fine. But, as with most things, there can be unscrupulous people who'll try and sell unreliable goods.

It is best to leave the delivery of the hives to your desired location to the seller. For him, this is a familiar thing. But if you're not used to transporting bees, things can go wrong and you might be faced with bees who've been inadvertently harmed or distressed.

In bee-related literature, you may read that the best time to acquire hives is the beginning of April when the willow blooms. But really the best time for delivery of the hives to the apiary is the second half of May. You can tell the way the bees fly, by the number of stuck frames, if you have a healthy and strong family or a weak and problematic one.

FOUR

Bee Care

Bee stings can be painful. Bees have no teeth, or at least not those that humans breed. Somewhere in the jungle, some bees live without stings, but they do have mandibles; they can bite humans with them. But such exotic bees are a long way from most people. Beekeeping does have the occupational hazard of stings and resultant pain. Neither tight clothing, nor protective nets, nor even rubber gloves will save you if you are dealing with an aggressive bee. Over time, your body will respond to bee venom by developing a stronger immune system. You will get sick less often and probably be more able to handle the occasional sting.

The main mistake of a novice beekeeper is to try to master everything in the first summer and to try and put the acquired knowledge constantly to the test. There are no miracles. Beekeeping, like any craft, takes more than one year to learn. Much of what you do will look clumsy and awkward at first. Do not despair. You will only get better!

What do smart people say? Trying to do something will inevitably mean some mistakes. The bees will teach you everything. The main thing is to love them and the business that you have become engaged in. The growth of expertise will be different for everyone. Between one and three years is often enough. Some people, unfortunately, may spend fifteen years and still won't gain any expertise; they will only ruin one colony after another. In short, it all depends on the personal qualities of people.

The most important thing for a beginner beekeeper is not to harm the bees. You are not a child, and a hive is not a toy. You don't have to dismantle it and probe around all the time. If you're trying to help the bees, there is no need to inspect the nests two or three times a week. You wouldn't want someone taking the roof off your home and looking down at you on a continuous basis; bees are no different. An overabundance of attention on your part is guaranteed to lead to a violation of the thermal regime in the nest, the distraction of bees from the work performed by them inside the hive, and a slowdown in the growth of the colony. The quality of your bee care does not depend on the number of examinations you carry out every month. When it is necessary to perform work inside the hive, do everything quickly and accurately.

It takes an experienced beekeeper about five minutes to inspect, for example, a bee colony that has not overwintered well. To carry out a spring check of a family with the performance of those works that will ensure the normal existence of the family for 3-4 weeks, ten minutes is enough. A novice beekeeper, if he gets confused, may take much longer.

For a novice beekeeper who has a shortage of time and does not have a mentor nearby, it would be best to look after bees according to a simple plan, gradually getting used to work and improving practical skills. Depending on the terrain and weather, the time for inspection of nests and work in the apiary is different, but the intent behind all such inspections is the same.

A simplified version of working with a bee colony

The first opening of the nest is a spring inspection of the colony. It is carried out from mid-April to early May. During its implementation, the general condition of the family and the presence of a queen are checked. If necessary, colonies are transplanted into clean hives, fed with sugar syrup, and insulated. The first inspection of a colony takes 6 to 10 minutes for each colony.

Close up of a frame filled with honey being put back in the beehive by the beekeeper

The second time you open the hive is in the second half of May. Your main task is to expand the nest with one store extension. At this time, you can raise several frames and, according to the amount of available brood, evaluate the quality of the queen bee. The maximum allowable time for this work per colony of bees is usually up to 7 minutes.

The third time you work on the apiary is in the second half of June. In the presence of the first productive honey collection, it is necessary to carry out a purification pumping out of honey from the store extension in front of the main honey collection. What if there will be a good harvest in July? Most likely, there will be little honey, but just on this day you can practice on the honey extractor, pumping out some fresh honey for yourself. A second store extension must be installed. It takes 5 to 7 minutes to work with the hive.

The time of your fourth visit to the apiary depends on the quality of the main honey collection and is carried out in early or mid-August. This is the time when honey is pumped out of the hives. It is advised to carry this out even before the end of the main honey collection, but in practice, due to the fact that not all the honeycombs are sealed, this does not always work, therefore the beekeeper himself determines the time for pumping out honey, removing excess bodies and removing unnecessary combs. This work takes 15–20 minutes.

For the fifth time, you open the hive in the second half of August, immediately after removing the store extensions, to determine the feed stock and, if necessary, to feed the bees. All work on the introduction of stimulating or replenishing winter stocks of dressings must be completed by August 25. This inspection takes 7 to 10 minutes to complete. For top dressing, 1-2 minutes.

In mid-September, you should carry out work to prevent varroosis by placing an acaricidal strip in the center of the hive.

Before inspecting the hives

1. A beginner beekeeper should think carefully about what he is going to do with the hive, and mentally imagine the whole course of work.

2. Take your time to prepare everything that may be required: inventory, hulls, feeding, spare hives. It all depends on the nature and volume of work performed.

3. Thoroughly wash your face and hands, do not over-use fragrant soap. When it's hot it's better to take a shower.

4. Correctly wear protective equipment.

5. Kindle a smoker with dry rot; they give a lot of smoke. And do not forget to take a stock of rot with you! You don't want to be there at your hive and have to go back to get more rot and rekindle the smoker.

The procedure for examining a bee colony

In order not to waste time and effort, bee colonies should be examined in a certain sequence:

1. Approach the hive from a certain side, from the side usually "out of the sun" for better visibility.

2. Let a few puffs of smoke into the entrance.

3. Carefully remove and set the cover obliquely to the back wall, with the roof to the hive.

4. Carefully remove the pillow and put it in the cover of the hive so that it does not get dirty.

5. Inspection should begin with the frame closest to you. Open the canvas with an energetic movement along the frames, but don't be too abrupt. If you remove it across, you can pull out or move to the side the frames firmly glued to it with propolis and accidentally crush the bees. Two or three puffs of smoke are quickly given along the open frames, the canvas is placed on the nest, tightly covering all the frames with it, so that the bees on top cannot leave the hive.

6. Not earlier than a minute later, begin to disassemble the nest and carry out the planned work. The beekeeper is located on the left side of the hive. Two or three extreme frames are advised to immediately pull out, inspect, and put in a portable box along with the bees sitting on them. The box is immediately closed with a cloth. This will create more space for viewing the hive.

7. In order to pull the frame out of the hive, carefully peel it off with an apiary chisel from the walls of the hive, firmly grab it by the shoulders and gently pull it up from the hive, trying not to crush the bees sitting on the walls of the hive and the side frames of the frame.

The inspected frame is held only above the hive so that the bees or the queen that have fallen off it fall not into the grass or the ground, but back into the hive.

If kept flat, the warm, soft comb can bend or, even worse, fall out under the weight of the honey. And during the honey picking period, in this position, a spray of honey can pour out of it and lumps of freshly brought pollen can fall out. During such work, it is best to use a spare canvas and immediately cover the examined streets with it. You should not open the nest entirely; you need to move the main lap as you examine the frames.

It is advisable not to forget about the smoke, and, from time to time, as soon as the bees start to run nervously around the frame or jump out of the streets, give them a little smoke. But everything depends on how excited the bees are. It's rare, but there can be occasions when the bees will behave very calmly and the extra smoke only makes them nervous. Usually, though, you do need a bit of smoke to sedate them.

Smoking bees

The checked frames are very carefully lowered into the nest and the next ones are also carefully removed from the hive. While you're doing this, don't touch the inner walls of the hive.

The nest is assembled in the reverse order.

Sometimes beekeepers have to inspect the family's nest, on which a store extension is already installed. Then, as always, smoke is first blown into the notch, then over the frames of the store. The store is removed and placed on a prepared table or, best of all, on the lid of the hive. Immediately smoke on the open frames of the nest and cover them with a piece of cloth prepared to the size of the hive.

Very often, only a partial inspection of the hive is required. For example, when you want to make sure that a young queen has begun to lay eggs. To do this, it is advised to immediately remove the frame, but not from the edge, but immediately from the center of the nest and, seeing the eggs laid on it or if they're already larvae, put it back, and close the hive again. If no egg-laying is found on the removed frame, then two or three more frames are looked at until you see the brood or make sure that the queen has not yet started laying.

Weather conditions when working with an open hive

The main requirements for weather conditions when working with an open hive are as follows:

- calm, sunny weather,
- the temperature in the shade is not lower than 15 degrees Celsius.

If the beekeeper has to work at a temperature significantly lower than the specified one, then when examining the hive, those frames that he does not hold in his hands need to be covered with a spare canvas to prevent hypothermia, and, when there is no honey collection - to prevent the bees from flying away.

A novice beekeeper must remember a couple of common truths. Firstly, work in the bee's nest cannot be interrupted until it is completed. You can't wander off and do something else half-way through. Focus on the inspection, finish it, and then allow the bees to go on with their lives. Secondly, any inspection of the bee colony is carried out only when necessary. Casual curiosity isn't a justification for opening the hive.

You cannot inspect the hive in rainy weather.

After each examination of the colony, the bees will stop working for 2–4 hours, and if this happens during honey collection, they lose a certain amount of honey. They have to spend time and effort fixing the damage caused by the awkward movements of the beekeeper, but even with careful work, the smallest interference in the life of the bees can be detrimental for them.

Different breeds of bees have varied dispositions of honey, bee bread, and brood. Therefore, when you're inspecting, the nesting combs should not be moved, so as not to create unnecessary work for the bees in restoring order.

FIVE

Tips for Beginner Beekeepers

Most of the troubles with beekeepers are due to inadequate knowledge of their subject or haste. To avoid or reduce unnecessary losses due to carelessness and wrong actions, beekeepers should acquire certain knowledge and, above all, study the biology of honey bees, as well as systematically familiarize themselves with periodicals in their specialty. Then, with a careful attitude to bees, you can learn to understand their "language" and needs, and you can achieve success.

Before purchasing bee colonies, you need to buy or make the necessary implements. You need beehives, face nets, an apiary chisel, smoker, drinkers (feeders), a portable box, spur, foundation, and beehive frames. Further, the need will inevitably arise for the purchase or manufacture of other beekeeping accessories - a honey extractor, a swarm, a workbox, a carpentry tool, spare hives, dividing grids, etc. But the main acquisition should be considered knowledge in your business - patience and the desire to master methods of caring for bees. They are extraordinary little creatures of nature. The more we can do to make them comfortable and secure, the better it is for us and our world.

How many families do you need to acquire at once?

A novice beekeeper should not be afraid to acquire several families at once. Having bought 2–4 families, he will be more active and bold in getting involved in work and acquiring skills in caring for bees. Having several families reduces the risk of being without them in case of erroneous actions and unforeseen circumstances (death of a queen in winter, bee attack, etc.). You won't have all your eggs in one basket. By purchasing several families, one can count not only on the successful mastering of the technology of bee care, but also on compensation for the costs associated with the purchase of bees.

Rules for inspection, feeding, and keeping bees

You need to work with bees confidently and boldly - but also carefully and calmly, without sudden movements and sounds. When inspecting the nest, you must not stand in front of the entrance and interfere with their flight. We must try not to create unnecessary noise and crush the bees. The smell of an accidentally crushed bee awakens the aggressiveness of its companions.

The beekeeper's clothing should be clean, light colors (white coat, gray overalls, etc.).

The basic rules for examining and feeding bees are as follows:

✓ Give top dressing (this is done in autumn and spring) in the evening, at least early in the morning.

✓ Before the inspection, it is necessary to ignite the smoker and bring the necessary tools and accessories to the hive to be inspected.

✓ The feeder should be installed in such a way that, during subsequent feeding, bees are excluded from flying into the space above the lap.

✓ Always have at least a week's interval between examinations of bees, and always in good (sunny and calm) weather.

✓ Bees need to be fumigated only if necessary. Abundant smoke irritates them.

✓ A very aggressive colony should be examined with an assistant. Before the inspection, 2-3 smoke puffs should be allowed into the entrance. After a minute, remove the roof, unscrew part of the canvas and put a puff of smoke on top of the open frames. If, during the inspection, the bees nevertheless show aggressiveness, the beekeeper needs to fumigate his hands;

✓ In case of severe irritation of the bees (this may be the result of an attack by thief bees or improper actions of the beekeeper), it is recommended to stop work and return to it after the colony calms down (usually 30-40 minutes is enough);

✓ During the honey-free period or cool time, the entire hive is not kept open. During the inspection, use a spare canvas, which is used to cover the examined frames (it is enough to keep 2-3 frames open).

✓ Frames removed from the hive (with and without bees) must be put in a portable box and closed with a lid;

✓ When inspecting the nest, it is necessary to have nearby dishes for collecting cut out drone brood, pieces of wax, etc.

✓ In the apiary, especially near the inspected hive, do not leave open honeycombs, dishes from syrup, knots, plug-in boards taken from the hive, and other accessories that smell of honey, including clothes that the beekeeper wore to carry out the inspection;

✓ In case of occurrence or threat of theft during the inspection of families, it is advisable to use an apiary tent.

If there is rather difficult work ahead (for example, re-forming the nest with the installation of dividing grids), it is advisable to first repeat to yourself (or write down on paper) the sequence of work and possible options. Prepare everything you need. A complete examination should be infrequent (2-3 times per season). Also, you should carry out the following in the winter - revision or transplantation into a new hive, selection of honey, determination of the quantity and quality of feed, as well as the formation of a nest. Some of these operations can be combined.

Air temperature during inspection with complete disassembly of the housing unit must not be lower than +16 ° C.

It must be remembered that any intervention by the owner of the hives disrupts the steady work of the bees. If they are doing well, they should not be "helped." The main concern of the beekeeper is to provide the bees with a work front. It is necessary to expand the nests in a timely manner with honeycombs. If the bees' flight performance has deteriorated, it is necessary to establish the cause and eliminate it.

Violation of the basic rules for examining and feeding bees can harm the normal development of the family. Accuracy in work and an attentive attitude to bees will eliminate stressful or harmful conditions in the hives.

Strong families

The beekeeper should strive to keep only strong families. It is better to have a few healthy and successful hives than a lot of poor ones. The main principles of having strong families are:

✓ Improving the quality of bee colonies through selection (the basis of success is a good queen);

✓ Culling of weak and sick families in the first year of life of the queens (after honey collection, and those affected by infectious diseases - immediately after the diagnosis is established);

✓ Leaving in the winter only strong and medium-strong families with a sufficient amount of good-quality food;

✓ Improvement and increase in the duration of honey collection due to migrations;

✓ Timely implementation of preventive work to prevent diseases. Oservance of labor hygiene in the apiary.

Beekeepers are guided by different technologies for keeping and breeding families, but the essence of these technologies is the same - year-round maintenance of strong families.

In this regard, replacing the queen in a weak colony will only partially improve it. The bees of a weakened colony are unlikely to be able to create optimal conditions in the nest that would allow the new queen to realize her potential.

The set of conditions conducive to an increase in the productivity of bee colonies and the growth of their strength includes the quality of the queen, the shape and volume of the bees' dwelling, the technology of caring for them, the quality and quantity of feed.

Feeding the bees

The main food of bees is honey and bee bread. The beekeeper must make sure that there is enough good quality food in the family all year round. Some apiary owners, when combining families at the beginning of the main honey collection, take all honey from bees in order to awaken their self-preservation instinct and increase the intensity of honey collection. This technique is unlikely to be successful for novice beekeepers. In addition, a complete selection of honey creates a stressful situation, which does not always pass without negative consequences for the bees.

Supplemental feeding of bees

In connection with all of the above, the beekeeper, when examining the nest, should pay attention to the stocks of honey and bee bread. If there is a lack of these substances, the keeper must feed the bees; that is especially important in early spring and in preparation for winter. With an abundance of honey, the bees work better and the colony thrives.

The other extreme should also be considered. In the winter in the nest, do not leave extra (not tightly occupied by bees) frames, especially full-fledged ones. If families spend the winter outdoors or in unheated rooms, honey in full honey frames, not occupied by bees, cools down. This causes harm. In the spring, such honey cools the nest, becomes inaccessible to bees, and creates conditions conducive to the development of nosematosis which is essentially bee dysentery.

Sanitary condition of the apiary

Beekeepers should systematically inspect the area in front of the hives and pay attention to what bees are throwing out of the hives, as well as observe their behavior and flight operations. If such work is disrupted or there is suspicious debris in front of the hive, this family should be given special attention. If the bees began to discard dry larvae, there is reason to assume that the colony is sick. Examine it, clarify the diagnosis, and start treatment, as well as think about the prevention of the disease in other nearby families. During the inspection, it is necessary to assess the condition of the brood and feed stocks.

It's an axiom of medicine that it's better to prevent disease than to cure it. We will briefly mention the most necessary sanitary and preventive rules, which are as follows:

✓ It is necessary to regularly disinfect the hives, inventory, tools, and clothing. If necessary (in case of infectious diseases), disinfect the territory of the apiary.

✓ The fore-hive areas should be kept clean (the garbage collected from them, including the corpses of bees, should be burned).

✓ In the case of illness of individual families, it's important to consult a veterinarian. Pathological material (samples - as a rule, 50-100 live bees from a diseased family, or, in case of a brood disease, a sample of a honeycomb with sick and dead larvae measuring 10 x 15 cm) should be sent to the nearest veterinary laboratory.

✓ Establish the cause of the family's illness and take measures to isolate it from other families.

✓ It is imperative to take the measures prescribed by the veterinarian to eliminate the disease and eliminate the causes of its occurrence, as well as prevent the disease reaching other families.

For disinfection of hives and small metal beekeeping equipment after mechanical cleaning from wax and propolis, use one of the following solutions (hot, temperature 50-70°C):

✓ 5% soda ash.

✓ 2% caustic soda.

✓ 4% causticized soda-potash mixture.

For the disinfection of winter huts, cell stores, nomadic booths, and storage rooms, you should whitewash walls and ceilings with 20% freshly slaked lime.

Dressing gowns, towels, and face nets, as well as oversleeves and other items of clothing should be boiled for 30 minutes or immersed in a 2% hydrogen peroxide solution for 3 hours, and then washed in water.

One method of disinfection of beehives and other wooden products is burning the inner surfaces slightly with a blowtorch. After burning the surface of the hives and other accessories, wipe them with a rag soaked in alcohol or a triple concentration disinfectant solution.

SIX

Be Successful - Disease Prevention, Swarming Prevention, Robber Bee Prevention Measures

Drinking water is of significant importance for the prevention of bee diseases. Providing the apiary with clean, slightly salty water will protect the bees from relying on impure water sources. On cold, spring days, it will lessen the chances of unnecessary fatalities.

Water for the bees

If you want to take care of your bees, you'll need to know the signs and methods of dealing with the most common diseases. We will tell you about the treatment and prevention of some common and dangerous problems.

Three serious bee diseases

First of all, mention should be made of varroosis - a disease of larvae, pupae, and adult bees, the causative agent of which is the varroa mite.

Varroosis symptoms include a decrease in the abdomen and shriveled wings. Without treatment, colonies with varroosis become weaker and die, especially after processing large amounts of sugar syrup in the fall or spring.

Currently, quite a lot of effective drugs and methods of combating varroa mite have been developed: bipin, phenothiazine, formic and oxalic acid, heat treatment, etc., but, from our point of view, the most rational, harmless, and convenient means of combating it are apistan and bayvarol.

A PVC plate impregnated with one of these preparations is placed in the center of the hive, passing it into one of the streets. In a strong family, 2 plates are placed in streets, 3–4 frames away from the edges of the nest. Plates are left in the nest for 25–35 days in spring or autumn immediately after honey is taken. In a summer broodless nest (period of swarming of bees), it is enough to leave the plates in the hive for only 3-5 days. Annual treatment ensures the well-being of the apiary for varroosis. Optimum results are obtained by simultaneous treatment of all apiary families.

The second extremely widespread disease of bees is nosematosis; the causative agent of the disease is the intracellular parasite Nosema apis. A clear manifestation of the disease is noticeable in late winter and spring. Bees will excrete liquid feces in the hive or near the entrance. The numbers of the colony will decline. An unpleasant odor begins to emanate from the hive since its walls and combs are covered with feces. Bees become lethargic, their abdomens are elongated, swollen, and soft.

The basis for prevention of nosematosis, like other diseases, is the observance of hygiene rules when servicing bees and the systematic renewal of combs so that the entire stock is renewed every three years, as well as storage in winter of usable combs in acetic acid vapors.

For the treatment of this disease, sugar syrup with fumagillin is fed to bees after pumping out honey or in early spring after they've been flying around. The content of one bottle of the drug is designed for 25 liters of syrup. Fumagillin is dissolved in warm water and added to warm syrup. It's best to do this immediately before distribution to the bees since the dissolved preparation quickly loses its healing properties. The syrup is given 5 times, 1 liter to each family every other day, or 0.5 liters daily for a long time - 23 weeks.

In the spring, fumagillin, dissolved in an appropriate proportion, can be given in a sugar-honey dough 2–3 weeks before the bees leave the hibernation house. In this case, 4 kg of powdered sugar, 1 kg of liquid honey, heated and dissolved in a small volume of water, and the contents of 1 bottle of the drug are kneaded. 1 kg of dough is enough for a bee colony for 3-4 weeks (1 bottle - for 5 colonies). If fumagillin is not available, trichopolum dissolved in syrup (0.5 g per strong family) can be used.

And finally, in recent years, ascospherosis, caused by pathogenic fungi of the genus Ascosphaera, has caused great harm to bees. This infectious disease affects the brood, primarily the drones. Infected larvae become soft, pasty, yellowish in color, and then harden, covered with white mycelium. Sometimes ascospherosis is called calcareous brood. Since the dried larvae seem to be liming, they are easily removed from the cells.

The bees throw them out of their combs to the bottom of the hive and on to the landing board. The discovery of dry larvae indicates a disease of the family. This means that it is necessary to immediately start treating it and, at the same time, stop it spreading to other hives. Severely affected brood combs are removed from a sick family, the nest is shortened, and good ventilation is provided.

The family needs to be transplanted into a new hive - a disinfected one. The honeycombs in the nest are gradually replaced with foundation and, at the same time, with an interval of 5–7 days, a medicinal syrup is given. This should be done until the dry larvae disappear. To prepare a medicinal syrup, sugar and water are taken in a 1: 1 ratio. After it cools down a little, a solution of 5% iodine (5 ml per 1 liter of syrup) is added to the warm syrup.

The medicine is distributed to families at the rate of 100–150 g per bee lane. This method of treating ascospherosis was developed at the Research Institute of Beekeeping and has proven itself in practice, surpassing other drugs in efficiency (for example, nystatin and unisan).

Swarming prevention care

Natural swarming of bees reduces the honey productivity of the apiary. Therefore, beekeepers should be aware of the signs that a bee colony is preparing to swarm and learn how to prevent it.

A bee swarm

Preparation for swarming begins long before the swarm takes to the air. The earliest manifestation of a family's preparation for swarming is intensive rearing (a sharp increase) of drone brood. Then, as the moment of swarming approaches, the number of signs of preparation for it develops. This, in particular, is the rebuilding of queen bowls, sowing them with eggs, growing queen larvae, sealing queen cells, and reducing the building and flight activity of bees. Additional signs of preparation for swarming are the termination of egg-laying by the queen and the grouping of bees in clubs in the subframe space, and sometimes at the arrival board. However, the latter sign may also indicate insufficient ventilation of the nest.

The following techniques can help reduce or prevent swarming:

✓ Selection and artificial reproduction of families from highly productive queens.

✓ Annual replacement of at least 50% of the queens (after 2 years of life, the queen becomes less productive).

✓ Shading the hives and providing them with good ventilation during the hot season.

If the beekeeper during the swarm season (late May - early July) does not have the opportunity to constantly be in the apiary, measures should be taken to stop swarming. However, there are times when the swarm cannot be suppressed. If a hive looks like it's moving toward a swarming state, it's useful to equip it with a special device that prevents the queen from flying out. The simplest such design would be a gateway through which worker bees can pass through freely, but the queen or drone cannot.

The main disadvantage of such minelayers is that they impair the ventilation of the nest, especially when some of the slots are clogged with drones. In hot weather, leaving such minelayers is unacceptable, especially for a long time.

If the beekeeper considers starting a new family from the swarm, then, having removed the veranda together with the mattress catcher and the bees sitting on them from their native hive, they must be transferred to a newly-formed hive. The bees are shaken off on to the frames, and the queen is released from the mattress trap. To do this, put it on the frame of the new nest and open the grate. The queen will rush into the nest by herself, and the remains of the swarm will follow.

Catching a swarm

Every year, a very large number of swarms are lost, either because the beekeeper with an apiary far from his home cannot exercise sufficient surveillance, or because he loses interest. The value of these swarms, lost to everyone, represents considerable capital and, cumulatively, a loss for the economy of the country.

Far too many people still imagine that trapping a swarm that has already left the hive is not within everyone's reach; some others believe that it takes a certain power, almost a gift. There are certainly many who would like to practice trapping, but they did not quite know how to do it.

To recover a swarm hanging from the branch of a tree, or on any other natural element, it is advisable to equip yourself with a swarm harvester.

Shaking a swarm off a branch

The swarm harvesters make the capture much smoother and less disturbing for the bees because, once the swarm has been captured, there is no need to shake it to knock the bees down. Simply open the zipper just above the beehive frames.

If you have a swarm harvester, it is usually easy when the swarms are hanging from branches. You will need to place your swarm harvester on a broom handle or a telescopic rod. It is then sufficient to open the jaw of the instrument so as to form a "sock" or "cuff," the lower part of which will be closed with a lace.

When we are ready, we only have to place the opening of the swarm collector under the swarm and "wrap" it completely. Of course, you have to be gentle - don't rush the maneuver. When the whole swarm is thus enveloped, you use a sharp blow to the branch to "unhook the swarm" so that it falls completely into the sock. Then all you have to do is pull on the jaw closure string to trap the swarm.

Now that the swarm is captured in the sock, it should immediately be placed on the frames of the hive or beehive. We then slowly open the zipper of the swarm harvester, and, quite naturally, the bees will be attracted by the frames of the hive that you present to them. They'll start to distribute themselves over the frames.

There are a number of reasons why it's important to catch the swarm:

✓ It's possible that honeybees aren't native to your area.

✓ Swarms can nest in your home and inflict property damage or annoy your neighbors, especially in metropolitan areas.

✓ You don't want to lose half of your colony, especially if you live in a location where the honey season is short. This would drastically diminish the amount of honey your bees can produce.

Basic techniques for stopping robber bees

Cases of attacks by robber bees can sometimes be provoked by a careless beekeeper. He might try to dry the frames by exposing them to the open air. Or, if there is no honey collection, robber bees can attack, destroy the queen, and transfer the honey to their hive. At the end of the robbery, when the queen is destroyed, the bees of the defeated colony begin to help the winners. The nest remains empty (with dry combs), and the thieves switch to other families.

The main sign of the beginning of theft is a fight at the entrance of the hive. Robber bees can be distinguished by their behavior. They sneak up to someone else's hive from the sides and bottom, and fly out of the hive swiftly. At the very beginning of the theft, it is easy to stop it, but if it turns into a full attack when a mass of struggling bees accumulates at the front wall of the hive, you need to act decisively, not forgetting at the same time to prevent an attack on neighboring families.

As a rule, first of all, weak, queenless, and sick families are attacked. The theft that has begun does not stop until nightfall, and is resumed in the morning. To stop bee theft, a number of techniques can be employed. Sometimes one measure is ineffective:

✓ It is necessary to immediately close or reduce the entrances in all families for 15–20 minutes to a size that ensures the passage of 1–2 bees;

✓ A piece of cellophane should be put on the boarding board of the hive that's under attack. A cloth moistened with kerosene should be fixed on it (for 1–2 hours), while the entrance is closed or reduced to 0.5 cm;

✓ You can try irrigating fighting bees periodically (with an interval of 5-10 minutes) with cold water. Or you can fumigate with smoke from a dry mushroom-raincoat or propolised cloth;

✓ If you see a fight at the entrance to the hive, cover the entrance with gauze or transparent film (raise it every 15–20 minutes to let in its bees).

If these measures aren't working? This means that a robbed hive, and a hive with thieving bees, if it is from its apiary, should be removed to a winter house or a cool dark room for 1-2 days, and a temporarily empty hive with grass with an unpleasant smell should be put in their place.

In order not to provoke the occurrence of bee theft, the following rules must be observed:

✓ During honey-free time (in summer weather), the entrances should be kept short, especially in weak families (layering and cores);

✓ It is not recommended to keep weak, hungry, and queenless families in the apiary;

✓ It is necessary to distribute the syrup for feeding only in the evening when the bees finish their flight work;

✓ In good, but honey-free weather, it is not advisable to leave the apiary for a long time;

✓ Inspection of families in such weather should be carried out with the help of an apiary tent or in the evening.

If the robber bees' family is your own and you notice the situation at the beginning of the theft, and not at its height, the simultaneous closure of all hives' entrances for 10-15 minutes can stop the robbery. In this case, the attention of the bees (including those belonging to the robber colony) switches to finding an opportunity to get into their own hive. They get stressed out and forget about attacking another hive. After the entrances are open, the usual order will reign in the apiary, but this must be done gradually.

SEVEN

Creative Attitude to Beekeeping

A creative attitude to the care of bees is of great importance for the development of a beekeeper. Sometimes, to a large extent, the work of beekeepers is facilitated by simple techniques and devices.

Metal wire and pliers for beehive frames

Template for punching (drilling) holes in the side strips of frames

The simplest template is a metal or sturdy wooden plate that measures the dimensions of a side plank. Four holes were drilled in it strictly along the longitudinal axis. The template can be used to prepare basic frames, as well as multi-body hive and half-frames, but it is better to make your own template for each type of frame. The length of the template for the frames of the multi-body hive is 220 mm, and for the half frames - 135 mm. Punching the holes aligns the template with the side strip, eliminating the need to mark each strip. It allows you to get centered holes, which, in the future, eliminates distortions of the frame when equipping it with wire (reinforcement).

A device for electric frame tightening

This device is a transformer with an output voltage in the range of 6-12 V, two wires with lugs and a wooden mold, on which a sheet of foundation and a reinforced frame are placed. The device allows you to reliably attach the foundation to the reinforced frame in 20-30 seconds. The process is as follows:

On the piece, a foundation is placed, cut so that its upper edge touches the upper frame. The side and lower edges do not reach the corresponding strips by 3-5 mm. The frame, laid on the foundation, is pressed with a transverse bar with a load so that the wire evenly touches the foundation. Having connected an electrical voltage to the ends of the wire, you need to carefully watch so that the heated wire does not cut through the foundation. As soon as the wire begins to sink into the foundation, one of the tips should be detached. After 2-3 seconds, the waxed frame can be removed from the piece.

If the wire tension is weak, or if it is pressed unevenly against the foundation, gaps may remain. In this case, pressing the wire to the foundation with one tip, the other tip is attached for 1-2 seconds to the section of the wire where the connection with the foundation ends.

The experience of doing this is acquired quickly. There are two points to pay attention to:

✓ The wire, when reinforcing the frames, should be stretched evenly.

✓ Before starting, you need to determine the time during which the wire has time to warm up before immersion in the foundation.

At the bottom, the mold has two strips 25 mm high. They are necessary so that the frame, laid on the foundation, hangs on the wire, as it were, and does not touch the table. Otherwise, it will not be possible to achieve uniform pressing of the wire to the foundation, and the whole process will be in jeopardy. The size of the piece corresponds to the frame 435 x 300 mm. To add frames of other sizes, you need to make patterns that fit into them.

A device for loading hives

This is a welded metal structure or a structure assembled from strong boards and a gantry-type trolley. It makes it possible to significantly facilitate the work of beekeepers when loading and unloading hives, especially in nomadic conditions, when it is difficult to find helpers when performing heavy work.

The use of dividing grids for separating the queen and limiting her oviposition during the main honey collection allows the work of the beekeeper to be significantly simplified when selecting honey frames. It also increases the honey productivity of the apiary. When installing the dividing grates, look to see there are no gaps between their edges and the walls of the hive through which the queen can rise into the second half.

Honey harvesting

A significant relief when pumping honey is provided by the use of an apiary fork instead of knives, especially in nomadic conditions. At the same time, there is no need to constantly keep hot water on hand because the cage contains significantly less honey.

The following tips can also help you pump out honey. Before opening the combs, they must be sorted so that equilibrium combs are in the honey extractor every time. As a rule, old (dark) combs are much heavier than new ones because they contain a lot of bee bread. They are also stronger and can rotate faster. It is advisable to pump out old cells separately from new ones.

Sometimes it is not possible to select frames of the same weight. In this case, the honey extractor should be secured to the floor or heavy deck with three braces or screws to stabilize it. That will prevent it from moving and swaying during fast rotation.

Radial honey extractors are convenient for pumping honey (including those with a horizontal rotor shaft), but they are more cumbersome and heavier than chordial ones. Such honey extractors are indispensable in large apiaries. When pumping out honey, all frames are placed in these honey extractors with the upper bars to the outside, and the lower bars to the central shaft. Their disposable capacity is up to 50 frames. There is no need to turn the frames in them (honey is pumped out simultaneously from both sides of the frames). If the honey is thick, or has had time to cool, the rotation speed should be reduced, and the pumping time should be increased to 15 minutes. Otherwise, the honeycomb may break.

In chordial and tangential extractors, honey should be pumped out with greater care. To ensure that the frames do not break, especially those containing newly-built full honeycombs, honey must be pumped out in three steps: first, at low speeds, about half of the honey is pumped out from one side of the frame, then, turning them 180 °, all the honey is pumped out on the other side. After that, the frames are turned over to their original position and the remaining honey is pumped out.

It is always advisable to set the honeycomb in chordial honey extractors so that, when rotating, the upper bar of the frame is at the back. The rotation speed in all cases must be increased slowly, up to about 120 rpm. Pumping honey at this speed lasts about one minute, then it is gradually reduced.

You will find it convenient to combine the pumping of honey with the selection of honey frames from the hives and the return of the pumped-out combs (instead of the withdrawn ones) for draining or continuing the honey collection. If the honey frames have been in cool storage for some time, they need to be warmed up in a room with an air temperature of + 25-35 ° C a day before pumping out, otherwise up to 20% of honey may remain in the honeycomb, and the risk of cracking and breakage during pumping will increase.

Dealing with ants

Ants are a big concern for beekeepers. There are a number of ways of dealing with them. You can lubricate the underhive stakes with grease to stop the ants ascending. You can place a film between the hive and the stand, the dimensions of which slightly exceed the area of the underhive stand. The most reliable one is to observe the rules of hygiene and cleanliness in the apiary. To do this, you need to carefully distribute the syrup without spilling it on canvas or on the ground. Syrup should not be allowed to stagnate in the feeders or be scattered around the hives and left on the ground with rot, food and other debris. If these measures are not enough, you need to sprinkle the place chosen by ants with bleach or freshly slaked lime. It is better to do this work in the fall when the flight work of the bees stops.

Ants are scared off by the smell of dill, so you can put several dill leaves along with the stems on a canvas in the hives or on under-hive stands. Ants are also repelled by the smell of tansy, so it's a good idea if that plant can be sown in an apiary.

Beekeeper working at apiary

Swarming bees

The departure of a swarm in an apiary is always exciting. At the same time, it is sometimes problematic to remove the swarm from the grafting site, especially if it is grafted high and at a great distance from the tree trunk. In this case, a swarm harvester may come in handy. It is advisable to have several interchangeable handles-poles of different lengths for ease of use.

Then, depending on the height of the swarm landing, it will be possible to remove the swarm neatly and without fuss, without risking falling with it from a stepladder or staircase.

Moving a hive

Moving a hive with bees has its own nuances. When bees are removed from the winter house, the hives must be placed in permanent places, but sometimes it is necessary to move the hive to a new site. This operation is performed as follows.

First of all, you need to turn the hive in the direction of the line along which it is planned to move. After about an hour, the hive can be moved to a distance of about 1 m. The bees will immediately start looking for their hive in the old place. In about two hours, they will get used to it (only a few will be looking for the "house"), and the hive can be moved another distance. It can be moved 10–12 m in a day. The work must be done in good flying weather.

Obtaining beeswax

An industrially manufactured wax melter is expensive, and many beekeepers use homemade steam or water wax melters. We offer a fairly simple water wax melter that allows you to melt wax from raw materials by 80–90%. This requires an enamel bucket, an electric or gas stove, two nylon stockings, and a wooden crusher. The raw wax is placed in a double nylon stocking, which is dipped into a bucket half-filled with water. You need soft water - rain or snow. Together with the raw wax, it should occupy no more than 2/3 of the bucket volume.

Cover the bucket with melted wax and water, wrap it well, and allow the contents to cool slowly. In this case, the main debris settles to the bottom, and its light particles stick to the lower layer of wax.

Place the bucket on the stove and heat the water to a boil. After 30–40 minutes of boiling, the boiled wax material remains in the stocking, and the melted wax ends up on the surface of the water. For a more complete extraction of the stockings with wax raw material during the boil 1-2 times, knead with the wooden crusher.

In order not to burn yourself, take out the stocking to remove the buckets from the stove. Then, gradually lifting the stocking, its contents must be squeezed out again with a pusher.

If it seems that some of the wax has not melted out, the operation can be repeated by changing the water (in a different bucket), but this is usually not required. Let us only note that there is no need to strive to immediately melt a lot of wax raw materials (0.6–1.0 kg is enough). The greater the excess of the weight of water over its weight, the better the wax is obtained.

Information absorption

The profession of a beekeeper isn't suitable for people who are hasty or rash. If you receive information about any useful and, at first glance, tempting innovation, do not race off to immediately apply it to your colonies. In this internet age, there is a lot of information about; some of it is excellent, some of it, regrettably, is poor. So, first of all, you need to find out in what conditions (climatic, ecological, etc.) this beekeeping innovation was tested. Then, if a decision is made to use it, the experiment should be carried out first on 1–2 families. If the result is positive, the innovation can be applied to other families.

In this case, it is advisable to first consult with an experienced beekeeper who can help make the right decision. It's a great idea to participate in beekeeping societies; you can learn from experienced people and specialists. This allows you to quickly obtain the necessary information and acquire the necessary skills for caring for bees. All sorts of issues, including new methods of dealing with existing and emerging diseases, are often discussed.

Beekeeper's diary

It is useful to monitor natural phenomena, to keep records of temperature changes. It is good to note the maximum and minimum temperatures during the day, at least for the period from February to June of each year. In the diary for recording air temperature changes, the most striking natural phenomena important for beekeepers should be noted. In particular, the beginning of flowering of coltsfoot and hazel, the arrival of wagtails, the beginning of flowering of raspberries, fireweed, and linden, as well as those phenomena that are significant and convenient landmarks. It enables you to record and plan the main milestones over a year.

If you're very technically inclined, you can use your observations to create graphs. Diagrams of changes in the daily temperatures during the period from December to March and the maximum temperatures for the period from March to July are of practical value.

To get abundant honey, beekeepers must grow strong families by the time of the highest productivity of the main honey plants for a given area. Observations of natural phenomena and changes in air temperature help them in organizing work to build up the strength of bee colonies at the right time.

Comparison of similar graphs for several years with each other and with the temperature graph of the current year makes it possible to reliably predict the most suitable time for setting hives for the first flyby of bees (including the very early one). It also helps in seeing when preparation of the paternal and maternal families begins, the formation of the first hives, taking into account the forecast conditions for mating of queens (a week after the emergence of queens from the queen cells, the temperature should be at least + 22-25 ° C). It is also easier to plan a trip and family vacation when you have such schedules in front of you.

Keeping a diary allows you to predict the weather with reliability close to the official data of weather forecasters, which is very important for nomadic conditions. It helps in understanding what kind of weather to expect, and even folk omens. For example, the following has been noticed by some beekeepers: if the icicles are long, the spring will be protracted, and the arrival of wagtails and larks foreshadows the imminent onset of warm weather.

Beekeeping journal

Brief records about the state of colonies, carried out and planned works, all help to discipline the beekeeper and provide him with invaluable assistance in improving the technology of bee care, as well as in breeding work. For this purpose, an apiary journal should be started. It notes everything that was done during the examination, division of families and their unification, disbandment, or the preparation of the paternal and maternal families.

The journal can be kept in any form, but good practice dictates that it is imperative to note the date, the number of the examined family, its condition (total number of combs and the approximate number of bees), the amount of brood, bee bread, honey and its condition, what has been done (given and withdrawn), what else needs to be done and when.

It can be convenient to set aside separate sheets in the apiary log for indications of the control hive, accounting for spring and autumn feeding, information on treatment, and other sanitary and preventive measures. After pumping out the honey, it is necessary to take stock; you can reflect on the honey productivity, the general temperament of each family, and, on the basis of this data, you can take into account the winter hardiness in the previous season, and outline measures for culling the worst families and other breeding work.

At the end of the active season and the assembly of nests in the winter, the results of the year should be analyzed and recorded in the log. The results of assembling in the winter, information about the strength of families, the stock of honey and bee bread, information on how many frames were assembled, and how ventilation was provided can be conveniently tabulated. The most important conclusions from the analysis of the year also need to be entered in the beekeeper's log.

EIGHT

Conditions for Successful Wintering of Bee Colonies

The art of the beekeeper is tested during the wintering of bees. In order for bee colonies to safely endure the coldest period of the year, you need to prepare for it in advance. The main measures for preparing families for winter are as follows:

✓ After the main honey collection (this is usually the second week of August), the bees have to be given time to fix the nest that's been disturbed by the beekeeper.

✓ The volume of the nest should correspond to the number of bees: there should be no extra frames in the winter nest, a "beard" in the frame space is permissible.

✓ The winter nest must be limited on both sides with side heaters and plug-in boards, next to which you put one full-honey frame, then one honey-burr (or in strong families, two). The center of the nest is filled with brood frames containing at least 1 kg of honey before feeding. Frames should be brown or light brown, but not black.

✓ Honeycombs with brood that are unsuitable for wintering bees should be placed on the edge of the nest and removed after the bees emerge.

✓ Side heaters should be installed during the final assembly and reduction of the nest.

✓ Feeding of families should start immediately after preliminary assembly of the nest and drainage of combs and finish by no later than September 5-10 while the brood is in the colonies, and the glands of young bees are functioning.

✓ In the nests, there should be no honeydew and rapidly crystallizing honey (such honey is removed and replaced with good quality from the stock or sugar syrup).

✓ During the autumn feeding, each family is given 6–8 kg of sugar in the form of a syrup. This is made by dissolving 1.5 kg of sugar in 1 liter of boiled water.

✓ After finishing feeding, you need to check the sufficiency of the honey supply, as well as the absence of open brood. If the family continues raising abundant brood in September, at the end of the month it is necessary to check the honey supply again at least in 3-4 central frames and replace the combs containing less than 1.5 kg of honey.

✓ If feed extensions (hulls) are given in winter, their combs must be filled with sealed honey to the bottom bar. Otherwise, the bees will not be able to move to them in cold weather; that's likely to result in deaths in the colony, perhaps even on a large scale.

✓ The winter nest should be well ventilated - 0.5–0.8 cm2 for each "street" in each entrance.

✓ It is necessary to ensure good ventilation of the winter house (the cross-section of vents should be chosen at the rate of 7–8 cm2 for each wintering family).

✓ In winter, bees should not be disturbed, including by feeding, until the time of the spring flight comes; earlier feeding may be justified if the colony is very low on food. In this case, you need to give 1 kg of sugar-honey dough, which will be enough for the family for 3-4 weeks.

✓ In the spring (after swarming), it is necessary to replace the pallets-tablets or bottoms. If there is mold on the bottom and walls of the hive with an inseparable bottom, the family must be transplanted into a clean, disinfected, and heated hive or the mold and dead bees must be removed sequentially from one half, then from the other half, of the hive.

To ensure the successful wintering of families, it is recommended to have 2-3 frames of good-quality honey in stock for each of them, at least until mid-September. The best for this is the honey collected by bees at the beginning of the main honey harvest. It crystallizes less and, as a rule, does not contain honeydew. Honeycombs filled at least 50% in the first third of the main honey harvest and sealed with wax are taken from the bees and stored at a temperature of + 22-24 ° C until the final formation of the winter nest. If there are no suitable storage conditions for these combs, you can leave them in your hives by sliding them to the edge and making the appropriate marks on them. At the final formation of the winter nest, these frames are placed in the center. The remaining frames with honey from the final assembly will come in handy in the spring.

Conservation of spare queens in winter

The most acute problem is the need for queens in early spring when there is nowhere to buy them. Therefore, experienced beekeepers leave spare queens in the winter within 10-30% of the total number of families. The last figure corresponds to the case when the beekeeper is about to organize early layering. You can store spare queens in different ways.

If the need is not pressing, that is, if early spring layering is not planned, in a 12- or 16-frame hive, you can place a queen nucleus behind a blank partition, a family of bees on 3-5 frames. The nest is assembled in the same way as the nest of a normal family. It should have full-fledged frames: one with bee bread, the rest with honey.

In the case of spring preservation of all queens, the queen nucleus is deposited in a separate hive and further used as an early layering.

If the need for queens is greater, the hive can be divided into three independent sections (to collect three nests with their own entrances in the fall).

Proper care

The well-being of the apiary is largely determined by the proper care of the bees. The health of the family is a harmonious balance between its strength, the number of brood, the quality of the queen, the conditions of keeping, and the abundance of nectar-bearing pastures. The task of the beekeeper is to maintain this harmony and reduce or eliminate possible stress factors on the bees. Such factors include poor or low-quality nutrition, excessive cold and heat, dampness and excessive dryness, hormonal imbalances, exposure to chemicals, strong electromagnetic radiation, physical vibration of the hive, sound vibrations, and unnecessary manipulations and experiments by the beekeeper on the colony.

Thus, a lack of protein in the feed when assembling a nest in winter can lead to premature wear and tear of the nurse bees in the spring, slow development of the family, and an increase in the likelihood of disease. And vice versa, an extra honey-bee frame left in the center of the nest in autumn sometimes leads to an overload of the intestines of bees in the club. This will negatively affect their wintering; it could even lead to the death of the family.

Lack of honey (less than 1.5 kg for two neighboring combs) can stress the family. Bees, going up in the club, will find that there is no honey, they will start to worry, and this will distress and confuse the whole family.

Stress for bees is itself a negative factor, and, in winter, this condition can lead to other negative consequences. Bees respond to almost any anxiety by increasing the temperature in the hive, and this entails additional food intake and an increase in the load on their intestine. During long wintering, this extra intake increases the likelihood of diarrhea even in healthy families. In addition, when the temperature in the hive rises, the club loosens, and after its decrease, some of the bees find themselves outside the club and may die.

The presence of honeydew in winter food can be attributed to factors that increase the risk of wintering success. Therefore, before starting the autumn feeding of the bees, it's necessary to find out if honeydew is in the honey. For this, a lime or alcohol reaction is carried out.

The simplest and most accessible method is using alcohol with a strength of 96%. First, 3-5 g of honey is taken from several combs of the investigated hive with a teaspoon, mainly from unsealed cells. Then, in a glass test tube or flask to one part of the honey taken for the sample, add an equal volume of distilled (in extreme cases, rain) water and 8 parts of ethyl alcohol. The mixture is shaken and allowed to stand for about a day. If, after this period, a sediment (flakes) appears at the bottom of the test tube, that means that honey is honeydew. The more sediment, the more honeydew.

Knowing the needs of bees for food, fresh air (especially in the second half of winter), quiet and darkening of the room in winter, and meeting the other needs we've already mentioned, as well as observing the basic rules for keeping and caring for bee colonies during the rest of the year are all important. Addressing them will help prevent the occurrence of many stressful situations.

The key to health, strength, and honey productivity of bee colonies and maintaining the optimism of beekeepers are accuracy in work, compliance with measures to exclude bee thievery, year-round provision of food, proper preparation for transportation in summer, and wintering in autumn.

An apiary in winter

Recommendations for neighbors of beekeepers, visitors, and nature lovers

✓ Bees demand respect for themselves. They do not like strong smells: alcohol, garlic, kerosene, cosmetics, etc. Therefore, if there are such smells, you need to calmly, but as quickly as possible, leave the apiary.

✓ Do not wave your arms, especially near the hives. It is not recommended that you make sudden, energetic movements in general.

✓Be careful not to interfere with the work of the bees. Let them get on with their lives, and they'll probably do the same for you.

✓ In the summer, in every bee colony, there are from 30 to 70 thousand bees, but there is no need to be afraid of the huge numbers. Each bee has its own job: they usually only sting people who prevent them from doing that job.

✓ The smell of a crushed bee triggers a defensive reaction in her companions. Bees are very sensitive. They immediately look for the "killer."

✓ It is better to cover your head with a hat or scarf. Bees often get tangled in hair, and their buzzing can attract the attention of their comrades.

✓ If you are stung by a bee, do not squeeze the wound with two fingers; if you do, poison will get deeper into the wound. The sting is removed with a scraping motion of a nail or a small knife.

✓ A sting wound attracts other bees by its smell and encourages them to attack. So, wipe it with a damp cloth, validol, or a rubbed leaf of mint, lemon balm, catnip, or parsley.

✓ Always have antihistamines (diphenhydramine, suprastin, tavegil, claritin) and validol in your medicine cabinet. The first is taken orally, and the second is used topically (externally).

✓ Ask your neighbor beekeeper to warn you in advance when he plans to work with bees, and to let you know if his bees are aggressive.

✓ If you rub the exposed areas of your body with mint, catnip, or lemon balm, the bees will not attack you, even if they are agitated.

When selecting honey, beekeepers are also advised to rub their brushes with lemon balm since bees will not voluntarily give up their goods, especially if the honey collection is over.

NINE

Inventory for Obtaining Honey and Useful By-Products

A widespread method of removing bees from combs is simple - it is shaking off and sweeping them with a brush or a goose feather using a smoker, but it is laborious. The bees can also get upset at such treatment, especially during the honey-free period.

In large beekeeping farms, special blowers are used to remove bees from combs. By adjusting the hose outlet, different air flow rates are achieved. The recommended speed is 6–7 m / s.

When removing bees by this method, the store extension, slightly tilted, is placed above the hive or on a special stand so that the combs are almost in a vertical position. After that, the "vacuum cleaner" is turned on and the bees are blown out of the inter-frames. Flying bees return to the hive, and the young fall on the plywood, with which they are transferred to the hive.

In order to remove bees in a calm fashion, repellents are also used, that is, substances whose smell repels insects. These include propionic acid, anhydride, benzaldehyde, and carbolic acid.

In this case, make an evaporative frame and cover the roof with it, removing the canvas and insulation from the frames. The frame, consisting of a lattice and several layers of material, is moistened with a repellent and covered with a film or a metal sheet, preferably aluminum, painted black. For wetting, it is enough to spray 15 cm3 of repellent on the fabric. The evaporator frame is left on a rooftop or on a store rack for about five minutes. This time is enough for the bees to be drugged. Some will leave, and some will sink to the lower slats of the frames from which they can be easily brushed off with a goose feather.

It should be noted that propionic anhydride and carbolic acid are ineffective on cold days without the sun. Benzaldehyde, on the other hand, works better in cloudy weather, so, on sunny days, it is recommended you cover the evaporating frame with a moistened white canvas. Note also that it is not recommended you use carbolic acid as a repellent if any other drug is available, as there is some evidence that it has carcinogenic properties.

On small amateur apiaries, you can use honey removers. Several parts are built into a plywood divider, which is installed between the magazine extension and the body. The working elements of the given remover are elastic plates. Bees can get from the hive into the removers' body only by passing through the cracks between them. Having passed the plate, the bees partially block the gap and cannot return.

The remover between the honey body (store) and the nest should stand for 24-48 hours. The holes in the honey building are closed at the same time. After a day or two, the bees almost completely leave the honey extension, and it can be removed. The disadvantage of such a remover is the long time it takes to remove the bees. In addition, if the superstructure is not clear enough, drones can get stuck between the plates, clogging the holes.

Removers of a second type are called tray removers. The principal detail, in this case, is the tray. It should connect the front part of the forward-shifted honey body with the upper entrance of the nest. At the same time, plywood is placed under the honey body (store), which completely overlaps the nest body. Bees isolated in the store in a few hours move into the nest building, and it becomes possible to select honey frames.

There are other types of removers. Another is knives for unsealing honeycombs. Unsealing (opening) honeycombs with this tool is a long and laborious process. However, they are still in demand in small apiaries.

When opening combs in this way, you must constantly have boiling water and two knives on hand. While the honeycomb is opened with one hot knife, the second knife is heated by keeping it in boiling water. Of course, the knives must also be sharp. When sharpening, it is recommended to use coarse sharpening stones. The knife can be likened to a file with many fine sharp teeth. The caps are cut off by moving the knife from top to bottom so that the beading does not stick to the honeycomb.

Beekeeping knives have been improved many times. Currently, along with them, steam and electric knives are also used, that is, those in which the blade is heated by steam and electric current, respectively. Electric and steam knives are available in many unconventional shapes. The heating increases the speed at which the honeycombs are unsealed. At the same time, labor productivity is high - an average of 10–20 seconds is spent on one hundred.

The disadvantages of steam and electric knives include the relative complexity of their configuration and the need for an energy source (power supply or steam source). In addition, when unsealing combs with such knives, a lot of honey - up to 15% - gets into the hatch.

A more convenient tool, especially in the field, is an apiary fork. It is a strong tinned steel plate, to which 16-18 thin long steel teeth are welded, equipped with a wooden or plastic handle. The diameter of the teeth at the base of the plate is 1.2–1.5 mm, the length is 4–4.5 cm, the distance between them is 4 mm.

When unsealing the honeycomb, the teeth are inserted under the wax caps of the cells, starting from the lower border of the sealed area. The movements of the fork should be light, rotational, and the plate should be in the plane of the wax caps.

In large apiaries, more productive devices are used to unseal combs - vibrating knives, brush needles, chains, and other mechanical devices. The main requirement for them is not to allow prolonged heating of honey to temperatures above 40–45 °C; this creates a risk of hydroxymethylfurfural formation in it.

To make the work of pulling combs as convenient as possible, tables or special frames with pallets are used.

A honeycomb pulling table is an oblong metal box with legs, over which a gable frame is installed, which serves as a support. The bottom of the box has a slight slope towards the drain. Above the bottom, at a height of 5-10 cm, a lattice is installed to collect the cut wax caps. There is a drain hole with a tap in the center of the bottom. The width of the table should correspond to the size of the frame, and the height of the drawer should correspond to its height (taking into account the lattice).

On one side of the table, unopened combs are inserted into the box (as in a beehive), and on the other, closer to the honey extractor, pulled-out ones. When unsealing the honeycomb installed on the frame, the cut caps fall on the grate and stay on it, and the honey flows down. Wax adhering to a knife or fork is scraped off against the inner edge of the box or pallet.

In amateur apiaries, a frame with a pallet is often used instead of a table. It can be metal or wood. A large baking sheet is used as a pallet into which the frame is installed. Honeycombs are placed on it at an angle of 20-30 ° to the vertical.

The honey extractor marked the beginning of a fundamentally new technology in caring for bees. Currently, honey extractors of various designs are used to pump out honey.

Their principle of operation is based on the use of centrifugal force. All existing designs can be divided into four types: radial honey extractors, chordial, tangential, and chordial-radial (universal). The name of the types indicates the location of the frames in the honey extractor: in the radial frames, they are placed along the radius. In the chordial ones - along the chord. In the tangential ones - at an angle to both the radius, and to the tank wall, and to the chord. In the universal packages of the frames, they are located along the chord and are radially oriented.

Honey in chordial and tangential honey extractors is pumped out in three steps. In other types of honey extractors, this is done simultaneously from both sides, that is, without manually turning the frames.

The most common in amateur beekeeping are chordial honey extractors. Their performance is low, but they are simple in design and not bulky. Chordial honey extractors are produced in various designs and capacities (from 2 to 6 frames). One pumping cycle lasts 2.5–3 minutes. Currently, there are chordial extractors with automatic frame turning to increase productivity, but they are more complex and have a larger tank.

Radial honey extractors have a tank with a larger diameter (up to 1.5 m), can hold up to 64 frames, and, as a rule, are equipped with an electric drive with a variable frame rotation speed (from 80 to 450 rpm). The processing time for one batch of frames can be up to fifteen minutes, but labor productivity when using radial honey extractors is higher, so they are widely used in large bee farms.

Tangential honey extractors are intermediate between radial and chordial.

Universal honey extractors have a horizontally positioned rotor and are charged with frame packs or single frames. In addition, they may have a mechanism for automatically loading and unloading frames.

It should be noted that the area occupied by a universal honey extractor is minimal; therefore, it is advisable to use it, first of all, when there is a shortage of space, for example, in mobile pavilions. "Cassettes" with frames in universal honey extractors are most often inserted from the side and fixed with special devices so that they do not fall out. In this regard, we note that for each frame size, it is necessary to install its own fixing devices. This is one of the main disadvantages inherent in universal honey extractors.

A honey extractor machine

Filters

Honey is purified from mechanical impurities, as a rule, using two-section filters. Both filter sections are made of tinned metal mesh. The upper one, with larger cells, has an almost flat bottom, while the lower one is a sphere. It is equipped with a retractable wire frame that's used to support the filter to be installed on the container. The two-section design allows for a longer run without rinsing and subsequent drying compared to a single-section filter.

With a large number of bee colonies, it is advisable to use a three-section submersible filter with a higher throughput. This filter consists of three parts: a body with two taps and two filter elements that perform coarse and fine purification of honey, respectively. The upper tap, located at some distance from the upper border of the filter, serves to drain the filtered honey during pumping out. The lower tap, located at the bottom of the filter, is opened to free the latter from honey at the end of the work. If you plan to continue working the next day, you do not need to pour out the honey from the filter at the end of the working day.

The immersion filter is fully used if its sections are covered with honey. Then the wax impurities float on the surface and do not clog the cells of the filter sections.

Devices for liquefying rigid honey

It is known that over time, liquid honey shrinks (crystallizes), that is, it becomes rigid. For packaging, or at the request of the customer, it sometimes needs to be transferred back to a liquid state.

The simplest device for liquefying honey is a large container with warm or heated water, into which cans of rigid honey are placed.

A prerequisite for its liquefaction is temperature control. It should not be higher than 45 ° C, since, with a further increase in temperature, honey loses its antibacterial properties. In addition, the content of hydroxymethylfurfural increases in it.

To liquefy honey, you can make a simple device yourself. It is a box, all sides of which are lined with heat-insulating material (polystyrene, foam rubber, etc.). In the center of the box, at its bottom, a holder for an incandescent light bulb is mounted. In order to simultaneously heat two 50 kg cans of honey, the box must measure 1 x 0.5 x 0.7 m. The cans are placed on wooden slats - then warm air will heat the bottoms too. With the help of a 60 watt light bulb, honey is liquefied under these conditions after 2–2.5 days. For a more even heating of honey in the corners of the box, you can place light reflectors or periodically (1-2 times a day) turn the cans by 180 °.

You can use other devices, but it is best to pack honey until it crystallizes. This crystallization is a unique biotechnological way by which it preserves itself. It enables its nutritional and healing properties to be maintained for a long time.

Stationary equipment for unsealing combs and pumping out honey

Large beekeeping farms must have a production building equipped with modern technological equipment. There is a honey storage room, a thermal hall, a compartment for unsealing and pumping out honey. In this department, the required number of automated devices and tables for unpacking honeycombs, honey extractors, honey sedimentation tanks with filters, the necessary transport and lifting means, as well as other auxiliary devices, are installed.

The core elements of such production buildings are technological lines for pumping out and packaging honey. The technological process itself is as follows. Honey frames (in hulls) are transported to the thermo-room, where the temperature is maintained at about 35 ° C. In 6-12 hours, honey in combs heats up to 25-30 ° C and becomes suitable for pumping out. Cases with warm honey frames are delivered to machines (or tables) for unpacking honeycombs. After opening the combs, the frames are placed in honey extractors, and the honey is pumped out. The evacuated frames are loaded into empty cases and sent back to the cell storage for further use. Honey from the honey extractor goes through mesh filters to a heated bath and is pumped to the honey settling ponds where the temperature is maintained at about 45 ° C. From the sedimentation tanks, it goes to the filling lines.

Royal jelly collection devices

Royal jelly is a product of processing bee bread and honey by young bees using the secretion of the pharyngeal glands. This is a biologically active food for the larvae of queen bees. It is healing - it has a beneficial effect on the human body: it increases resistance to infections, improves vision, benefits the cardiovascular and nervous system, etc. In the last twenty or thirty years, medical and scientific attention to royal jelly has intensified, and many beekeepers have begun to extract it.

The devices for collecting royal jelly are in many ways similar to the devices used for raising queens and inoculating the larvae.

A bee cell with royal jelly in the wax comb

The main structure for collecting royal jelly is the graft frame. In fact, it is similar to the grafting frame for raising queen bees. The larvae are also grafted here, but the bees are only allowed to raise them for three days. After this period, the grafting frame is removed from the nest, the larvae are removed, the royal jelly is collected, and the frame is used for a repeated cycle. It has been established that most of the royal jelly is found in larvae, whose age is equal to 72 hours, so its selection should not be delayed longer than three days of rearing the larvae. If two grafting frames are placed in a family at the same time, a frame with a sealed brood is placed between them. In this case, the percentage of larvae intake increases. On each frame, three grafting slats are usually attached.

To collect royal jelly, dark-colored glass jars with a ground-in cork are used as the jelly quickly deteriorates under the influence of sunlight, moisture, heat, and oxygen. To avoid this, royal jelly is preserved with honey by mixing one part of it with 20 parts of honey.

This valuable beekeeping product can also be preserved in a 1% concentration.

At a temperature of -6 ° C, pure royal jelly in hermetically sealed dark glass jars can be stored for no more than six months, while at + 6 ° C, this can be done for only one day. In this regard, freshly extracted royal jelly should be placed in a refrigerator with a temperature of 6–8 ° C after 1.5–2 hours.

Devices for melting wax raw materials

Wax is a waste product of bees, produced by their wax glands. It is an irreplaceable building material for a bee's nest. People also use it, in particular in medicine, electrical and radio engineering, electroplating, and printing. Fresh wax contains 24 esters, 12 free and bound acids, saturated hydrocarbons, vitamin A, minerals, etc.

For novice beekeepers, wax is a scarce material, so they have to purchase the necessary equipment or make their own devices for melting wax raw materials. The simplest of these devices is the solar wax melter - an obliquely installed rectangular box, which has a tray (baking sheet) made of tinned tinplate and a trough for collecting molten wax.

To create a high temperature in the wax melter, double glazing is made in its lid. The lid should fit snugly to the walls, so a strip of cloth or foam rubber is glued around its perimeter. The walls of the wax furnace are painted black inside and out.

In a correctly mounted and installed wax melter, temperatures can reach 100 ° C.

The wax melter is installed in a sunny place - it must be continuously illuminated by the sun throughout the day. Periodically, the wax melter needs to be turned: for this, you can make an automatic turner. The mirrored reflector also contributes to the temperature rise in the wax melter.

The unit is loaded with raw wax material in the morning, spreading it out in a thin layer. At the same time, the hardened wax is removed from the wax melter. Water should not be added to the trough; it evaporates and will cause the glasses to fog up and the process of heating the wax raw material to deteriorate.

In the summer, the solar wax melter will relieve beekeepers from the worry of processing wax waste, which quickly deteriorates. In addition, the wax obtained in it is of high quality.

The disadvantage of the solar wax melter is that quite a lot of wax remains in the furnaces, especially when the old combs are overheated (up to 50%).

But with large volumes of processing of wax raw materials, it is more convenient to use industrially manufactured wax pots and press waxers.

Devices for listening to bee colonies

In winter, the main form of checking the state of bee colonies is acoustic control. It does not bother them and at the same time allows you to determine whether families are wintering well. For listening, you can use rubber or metal tubes, medical or beekeeping phonendoscopes, microphones with amplifiers, and other devices. It is recommended you listen to families when visiting the apiary, starting from mid-January. That is, of course, if you haven't noticed warning signs before that time of year.

The hum of a bee colony will allow you to get an idea of the state of its wintering if you know the following:

1). Quiet, barely audible sounds with an admixture of low tones indicate a good wintering state of the family.

2.) Increased hum indicates stuffiness in the hive, insufficient ventilation of the nest, and, possibly, thirst experienced by the bees. This happens when the entrances are narrowed, and there is an insufficient supply of fresh air for the colony, especially in slushy winters with frequent thaws. In this case, it is necessary to clean and widen the tap holes or unscrew the back corners of the hive.

3.) A discordant, uneven rumble, high tones of high intensity mean the possible death of the queen. This colony will need to be given special attention, and if the worst assumptions are confirmed, it can be merged with another family.

4.) Barely audible noise, like rustling leaves, means that the family is in danger of hunger. The bees need to be given sugar and honey dough.

5.) Strong and intermittent hum may indicate the intrusion of a mouse into the hive.

6.) The absence of any sounds obviously portends the possible death of the family. If, after lightly tapping your finger on the hive, the situation does not change, you can unscrew the canvas and establish whether the family is still alive.

The rubber tube makes it easy to listen to the bee colony if there is no strong background noise. The tube should have a smooth inner surface, 60-100 cm long and 5-10 mm in inner diameter. One end of it, which is inserted into the ear, can be cut into a cone if the tube is thick-walled. The second end is attached to the open taphole or pushed into it by 2-3 cm.

Similarly, families are listened to with a steel double-curved tube. It should have parameters close to those indicated for the rubber tube.

Medical phonendoscopes are more convenient than tubes because they free the beekeeper's hands. A flat funnel, attached to the tap hole, allows you to more clearly catch the hum of bees, and the headphones not only enable your hands to be free but also reduce the penetration of extraneous noise.

TEN

Frequently Asked Questions About Beekeeping

Beekeeping is becoming an increasingly popular pastime. Some questions on the subject reoccur - these are a few that deserve to be addressed before taking your initial steps in this field.

1.) *What are the fundamentals?*

The first thing that anyone considering entering this sector asks is, "What is beekeeping?" It is an agricultural operation that entails rearing bees with the purpose of gathering honey.

2.) *Is this a difficult job to perform?*

This is a hobby that requires initial motivation. Some people are fascinated by bees from the outset, while others become more interested as they learn about them. It's often followed by training in schools and with professionals. Or by reading a book such as this one. We also cannot ignore the project's financial implications. Indeed, you must budget for and purchase the appropriate equipment (overalls, jars to store honey, gloves, a beehive, a honey extractor, etc.).

3.) *How long do bees live?*

A colony consists of between 30,000 and 80,000 bees. The queen can live up to five years, while the workers live just thirty-five days in the spring and summer, and four months in the fall and winter. Males who are only present in the colony during the spring and summer would live for four or five months.

4.) *How does a bee produce honey?*

The work in the hive is evenly distributed: worker bees collect nectar and transport it to the hive, other workers distribute the nectar within the hive, and salivation transforms the honey nectar.

5.) *How many species of bee are there?*

There are almost 20,000 species worldwide, although not all are "exploited." Just a handful are bred, including Apis mellifera mellifera, black bees, Apis mellifera Carnica, and Apis mellifera Caucasica.

. . .

6.) *When a jar of honey is labeled "organic", was the honey foraged in an organic environment?*

No, not necessarily, because organic honey is defined as honey produced by bees that have not been treated with antibiotics.

7.) *What diseases do bees suffer from?*

When beekeeping becomes your primary hobby, you must become knowledgeable about bees and everything that affects them. The issue that frequently forms the fear of some aspiring beekeepers is "what are the most prevalent bee diseases?" Several diseases exist, and they are capable of wiping out an entire colony if preventative measures are not performed in time. Chapter Six discusses bee diseases in detail and also shows you preventative measures to keep your apiary disease-free. It is vital to have knowledge about hive and bee treatment in order to prevent or deal with such situations.

8.) *Why does honey come in many colors, scents, flavors, and how are they different?*

Real honey collected from wild bees, or domesticated honey bees, has no difference in composition and the nutritional values are similar, but the properties, color, smell, taste, and crystallization differ according to the type of flower the bees collected their nectar from.

- Honey from longan flowers is honey that smells sweet. It's the most popular, and it can be stored for a long time. It does not change color and does not crystallize.
- Honey from wild flowers is fragrant honey. When new, it is usually light yellow. It gradually turns darker over time. If stored for a long time, it may crystallize.
- Honey from sunflowers is honey that has a distinctive aroma and a pale yellow color. It is the easiest to crystallize. It's allowed to crystallize to be used as honey

cream or honey jam. This type of honey is favored by some people as it can be used as a spread in bread.

- Honey from sesame flowers is honey that smells like honey from longan flowers. You can leave it for a long time to crystallize.

9.) *What can I do with a jar of honey that has crystallized?*

Crystallized honey has not lost its nutritional qualities or flavor. To restore its creaminess, simply place the jar in hot water until it becomes completely liquid again. Or, as noted in number (8) above, you can use it as a spread on your pastries.

10.) *What is the difference between wild honey and farmed honey?*

It's no different because, in both cases, the bees are allowed to collect nectar naturally. Wild honey cannot be distinguished from farmed honey. Besides, nowadays it's very hard to find wild honey because the forests are steadily decreasing.

However, as a specialty apiary, you can set up an orchard (a longan garden in March or a sunflower field during the month of October, etc.). Honey bees naturally go to places where they can find the most nectar.

11.) *What can honey be used for?*

It can be used as a substitute for sugar in tea, coffee, Ovaltine, or in beverages such as honey and lemon. It can be used for cooking, such as marinating meat, pork, or making salad dressings. It is used to make desserts such as toast topping, pancakes, etc.

Additionally, it is used to make cosmetics such as hair treatments and face paint. It is also highly prized in the herbal industry as it is an essential component in making some vegetable smoothies, juices, etc.

Or simply eat it raw. Eating honey helps nourish the body.

12.) *What is bee pollen?*

Bee pollen is pollen that is attached to the legs of bees. Beekeepers harvest pollen from bees by the use of an extra-large comb that is kept in the doorway of their hives. When bees fly past it, the pollen is knocked off their legs and collected in a collection bucket below. This is then sold to be processed and sold as a dietary supplement.

13.) *Who should choose bee pollen?*

- People who have frequent colds.
- People who want to maintain healthy nails, hair, and skin.
- People who work hard, who have to use their brain for a long time, or are exposed to stress.
- Children who do not like to eat vegetables.
- Adults who want to avoid meat, milk, eggs.

14.) *Does bee pollen have any health benefits?*

Bee pollen has the following health benefits:

- Helps the body to be strong, refreshed, and rejuvenated.
- Helps treat allergies, high blood pressure.
- Nourishes the hair and the skin.

Note: Because it is a dietary supplement, it should be eaten continuously to see results.

15.) *What is royal jelly?*

Royal jelly is a food that the worker bees produce for the queen bee. This is the queen bee's food. It is a creamy white liquid. In royal jelly, there are many nutrients such as carbohydrates, proteins, and many vitamins.

. . .

16.) *What are the benefits of royal jelly?*

Royal jelly has the following benefits:

- Nourishes the brain.
- Strengthens the body.
- Helps you feel refreshed.
- Relieves fatigue.

17.) *What are the differences between bees and wasps?*

It's summer and you are making the most of your land when all of a sudden, a little hum is heard… and you find a "honeycomb" nest under your porch, in a tree, or by the pool. It could very well be that it belongs to wasps and not bees. Many people confuse bees and wasps. Certainly, at first glance, these insects can look alike. Here are some ways to recognize them:

Bees

A bee's body is stocky, hairy and its abdomen lined with black bands. Depending on the species, the space between its stripes can range from dark brown to yellow. It measures approximately 12 mm.

A group of bees leaving a hive form a swarm; it is made up of several thousand bees. Bees build a hive filled with honey. Bees are not attracted to our food and rarely sting. A bee dies after having stung; her sting is provided with a hook that remains in the skin.

Wasps

These insects are less hairy than bees. They have a bright yellow color with very marked black lines and an abdomen clearly distinct from the thorax (hence the expression "having a wasp waist"). They can measure between 10 and 25 mm.

Wasps make a paper pulp nest by chewing wood fibers which, mixed with their saliva, form a paste. Wasp nests can be visible, hidden underground, or in a wall. Unlike bees, wasps are attracted to our food and are aggressive. They do not hesitate to sting and can do it several times.

Have you spotted a nest of wasps or bees? Do not approach it, do not disturb it, rather keep an eye out to find out what is going on there.

Early in spring, the young queen leaves her shelter and goes in search of a suitable site for her colony. Once the site is found, she builds the nest and begins to lay eggs. The nest, at this stage, is small and can be removed quite easily.

During summer, the population increases ... and so do the comings and goings: as the larvae become workers, they emerge from the nest in order to find food for the new larvae. Eliminating the nest is now difficult and riskier.

If you see a nest of wasps or bees in the fall, be relaxed; this season marks the end of their life cycle, except for the young queens who will perpetuate the species. But no doubt they will soon leave to take refuge in a winter shelter.

Warning! Removing a nest yourself, or blocking the insects' entrances always involves risks. Do so carefully. Better still, leave this task to extermination professionals.

18.) *How can you prevent yourself from being stung by bees?*

Prevention

These insects sting for self-defense. If a bee bothers you, avoid any sudden movement. If it lands on you, let it go on its own or gently push it away. If you've disturbed a nest, get away as quickly as possible.

In nature

Keep sugary foods in closed containers: the adult bee looks for sugar/nectar to feed itself and it must bring pollen to the larvae of the colony. Avoid fragrant smells on your clothing. Watch your food carefully, especially that of children. Before taking a bite or sip of juice or sweet liquor, make sure there is no bee in it. Even if it's in your mouth or throat, a bee can sting.

Prefer long clothes. Tie up your hair. Wear a mosquito net hat if necessary. Avoid walking barefoot: some nests are close to the ground. If a bee enters the car, open the windows. It will come out on its own.

In the house

- Use garbage cans with lids.
- Do not leave any table waste lying around.
- Make sure all the windows in your house have screens.

As an apiarist

This has been discussed in detail in Chapter Three. A smoker is an important tool for an apiarist and will allow you to do your job without the bees bothering you too much.

19.) *What should be done in the event of a bee sting?*

Examine the stung area. In most people, there will be redness, pain, and swelling around the site of the sting. This reaction is normal, and although it may seem very worrying, it will go away within hours or days.

Just apply cold compresses and take pain control medicine (acetaminophen). However, if the local reactions are very widespread or accompanied by fever or local infection, consult a doctor.

20.) *Should I worry if I get stung?*

There are a few rare cases when you need to worry after being stung by bees. In these cases, the reaction becomes abnormal and worrisome as it occurs elsewhere than at the site of the sting. Signs of a serious reaction are: swelling of the face, generalized redness of skin, change in voice, difficulty swallowing or breathing, weakness, persistent vomiting, loss of consciousness or shock. Such a reaction can occur just a few minutes after the bite and be life-threatening. If one or more of these symptoms are present, act quickly: administer EpiPen® (available from pharmacies without a prescription), take an antihistamine (Benadryl®), and go to the nearest emergency department. If necessary, call for paramedics.

These symptoms are often present in people that are allergic to bee stings. People who have had an allergic reaction to a bee sting have a high risk of having another similar one or worse if stung again by the same species. Have you ever observed one or more signs of allergy following an insect bite? Talk to your doctor if you have such concerns.

If you have ever experienced strong reactions to an insect bite or a bee sting, see an allergist.

Bear in mind, though, these reactions are rare. The vast majority of people who've had a bee sting only experience some small-scale pain and discomfort, and it's usually easily treatable.

ELEVEN

Great Honey Recipes

Honey is a wonderful product - delicious, aromatic, and healthy. Of course, there are allergic reactions to honey in a few people, but this is no reason to deny a yummy treat to most people. Honey can not only be eaten with a spoon and put into tea, it can also be used to prepare all kinds of sauces, cakes, pies, drinks, desserts, and glazes. And what is most interesting - in many dishes honey does not lose its healing qualities! The main thing is not to heat it above 40-42 ° C, otherwise it will first become a very fragrant "sugar", and then it will become just sugar. Sounds complicated? In fact, cooking delicious and healthy dishes with honey is quite a simple task. You just need to master a few correct recipes!

A honey-based recipe

Honey fruit and berry wines

Only ripe, undamaged fruits are suitable for the preparation of wines. All berries (raspberries, strawberries and blackberries), as well as fruits, must be washed. For the preparation of most wines, it is recommended to squeeze the juice in advance, but you can filter the already fermented drinks, with the exception of plum wine, since plum pits can give it a bitter taste. It is not forbidden to blend (mix) juices before fermentation; if you do, the wine will have an original taste.

Raspberry honey wine

10 liters of raspberry juice, 5-6 liters of honey, and 15 liters of water are required. After mixing and adding a small amount of yeast, the dishes with wine are covered, and the fermentation process is monitored from time to time. It usually ends in 5-6 days. Then, after removing the wine from the sediment, the drink is transferred to the cellar. The wine is considered mature in 1.5-2 years.

Similarly, you can prepare honey-berry wines from strawberries, blackberries, currants, cherries, and other berries. Wine from pears and apples is prepared in a slightly different way.

Honey apple-pear wine

From pure pear juice with honey well-fed into the mix, the wine turns out to be sweetish and fresh-tasting. It will be more delicious if pear juice is mixed in equal parts with apple juice. The following recipe is recommended. Five parts of pear-apple juice are added to 10 parts of honey.

After fermentation, clarification, and maturation, a wine of excellent taste is obtained, but not earlier than six months later.

Sauces

Applesauce

6 apples, 1/2 cup honey, 1/2 cup water, 1/2 lemon juice, 3 cloves. The apples are peeled and cut into quarters. Pour water and lemon juice into a bowl, add honey, add cloves, and bring to a boil. Put the apples in and stew until the sauce is smooth, after which the cloves are taken out. Serve the sauce with chicken or roast pork.

Salad dressing sauce

1/4 cup honey, 2/5 cup milk, 1/4 cup lemon juice, 1/2 teaspoon salt, 1/4 teaspoon sweet red pepper. All sauce ingredients are mixed to make the mass homogeneous. Put in the refrigerator to thicken. This dressing is especially good for green salad or fresh cabbage salad.

French sauce with honey

1/2 cup vegetable oil, 1/2 cup lemon juice, 1/2 cup honey, 1/2 teaspoon hot red pepper, 1/2 teaspoon salt, 1 clove of crushed garlic. Put all the ingredients in a jar, close it tightly and mix thoroughly. You can serve French sauce with almost any salad.

Meat sauce

1 cup tomato juice, 1/4 cup vinegar, 1/4 cup honey, 2 cloves of crushed garlic, a little hot red pepper, 1/4 teaspoon salt, ground black pepper to taste. All ingredients are placed in a saucepan, mixed, covered and cooked for 10 minutes.

Apple sauce for beef

200 g apples, 100 g sour cream, 10 g flour, 20 g butter, 1 teaspoon honey, salt, parsley. Rub the peeled apples on a grater and cook in water for 10 minutes.

Grind sour cream with flour, honey, and 1 tablespoon of water. Add salt and butter to taste, sprinkle with green parsley and let it boil.

Chocolate sauce with honey

3 cups chocolate chips, 1/2 cup cream, 4 tablespoons honey. The chocolate is softened and cream and honey are added to it. This sauce is served with desserts and ice cream.

Second courses

Roast in marinade

1.5 kg of meat, 3 tablespoons of honey, 2 tablespoons of vinegar, 1.5 tablespoons of crushed garlic in a small amount of water, 3/4 cup vegetable oil, finely chopped green onions.

Mix honey, vinegar, garlic, vegetable oil, and the chopped onions thoroughly. Fat and films are removed from the meat, after which it is cut diagonally in the form of rhombuses. The meat prepared in this way is placed in porcelain dishes and allowed to marinate (it is left for 4 hours at room temperature or put in the refrigerator overnight). Then the meat is boiled in a marinade, then fried over medium heat for 6 minutes on each side. When serving, the roast is cut into portions.

Chicken with honey

Large chicken (1.5 kg), 1/3 cup honey, 1/3 cup flour, 1/4 cup lemon juice, 1 teaspoon ground red pepper, 1 teaspoon salt, butter. The chicken is cut into portions, doused in flour and fried in butter, then laid out on a baking sheet, salted, sprinkled with ground red pepper, poured with a mixture of honey and lemon juice, and baked in the oven for 15 minutes.

Pork chops with honey

6 pork chops, 6 tablespoons of honey, a cup of savory ketchup, lemon. Ketchup is mixed with honey, and the mixture is poured on the chops. They're then baked in the oven. When serving, place a slice of lemon on each chop. Chicken chops can also be made with this sauce.

Roast pork with sauce

2 pork chops, 3 tablespoons honey, 1/2 cup vegetable oil, a few cloves of crushed garlic, 1 tablespoon dry mustard. Vegetable oil, garlic, honey, and mustard are mixed and ground properly. This is poured on the chops, and they're put in the refrigerator for a day. The next day, the pork is baked in the oven for 30–45 minutes. Fried potatoes are great as a side dish.

Meat in prune and honey sauce

80 g of meat, 10 g of honey, 15 g of pitted prunes, 5 g of fat, 5 g of onion, spices and salt - the recipe is designed for 1 serving. The meat is fried in a pan, sautéed onions, prunes are added and simmered until tender. Honey is added to it before serving.

It is recommended you serve this dish in portioned pans, with the sauce in which the meat was stewed. The best side dish for it is stewed vegetables.

Lamb with mint and honey

4 chops, 1/2 cup honey, 1/4 cup water, 1 tablespoon vinegar, 2 tablespoons freshly chopped mint or 1 tablespoon dried, salt, black pepper to taste. All sauce ingredients are mixed and simmered under a lid for 10 minutes. Add salt, pepper, and fry for 5 minutes on one side, turn over, pour half of the mint sauce, fry for another 5 minutes, and pour the remaining sauce into the pan.

Goose with apples and honey

Medium goose, 15 small sour apples, 1 tablespoon honey, 4 tablespoons raisins, 8-10 olives, salt to taste, parsley. The prepared gutted goose is rubbed with salt on the outside and from the inside, 6–7 apples are peeled, cut into quarters and cored. The raisins are washed and soaked in hot water. After it swells a little, the raisins are mixed with apples and honey, stuffed in the goose and sewn up. The carcass is placed on a baking sheet with the back facing down. A little water is poured and placed in a preheated oven to a high temperature. As soon as the goose is browned, it is turned over with the cut down.

When the whole carcass is browned, the temperature is reduced, and fat that has melted from it is poured on it every 7-10 minutes. The goose will be ready in about 3 hours. A few minutes before the expiration of this period, place the remaining apples around the goose and pour more fat on it. The threads are removed from the finished poultry. It's put on a dish and garnished with baked apples.

Lamb stewed with honey

120 g lamb, 20 g honey, 7 g ghee, 5 g tomato puree, 3 g wheat flour, spices, salt to taste - the recipe is designed for 1 serving. Cut the lamb into 2 or 3 pieces, sprinkle with salt and pepper, fry, transfer to a saucepan, put honey, tomato paste, and spices (coriander, cinnamon, cloves) in it, add a little water or broth, and stew until tender. It is recommended you serve it with crumbly rice or boiled beans for a side dish.

Duck with honey

Large duck (2.5 kg), 1/2 cup honey, 1/3 cup orange liqueur, lemon juice, 1.5 teaspoons dry mustard, 5 lemon slices, 5 onion rings, salt, ground red pepper to taste. A prepared duck is pierced with a fork to drain the fat. After mixing salt, lemon juice, and red pepper, rub the carcass and put it in the oven to bake for 15 minutes at maximum temperature. After that, the temperature is reduced, and the duck is baked for another hour, draining the fat from the baking sheet if necessary. The honey, liqueur, and mustard are mixed and rubbed over the browned duck. The bird is decorated with lemon slices and onion rings and left in the oven for another 15 minutes, after which it is served.

Honey-marinated cutlets

8 lamb cutlets. Marinade: 1/2 part honey, 1/2 part apple juice, 1 teaspoon mustard, 1 teaspoon soy sauce, salt. The cutlets have the marinade poured on them, and they're left for several hours in a cool place. After this, they are fried on a wire rack in the oven at a temperature of 250 ° C. It is recommended you serve fried potatoes, green peas, and cabbage salad as a side dish.

500 g pork ham, a little vegetable oil. Marinade: 1/3 part honey, 1/2 part soy sauce, 1 tablespoon sugar, 1 clove of garlic, 1/2 teaspoon ginger. The meat is cut into thin pieces, coated with marinade on both sides and left in a cool place for 6-8 hours. After this time, the pork is fried in a pan. Served with crumbly rice, canned corn cobs, bell peppers, and onion rings.

Pasta in milk with honey

125 g pasta, 200 g honey, 1 tablespoon raisins, 1/2 cup water, 2 tablespoons butter, 2 tablespoons chopped nuts, 2 cups milk, teaspoon vanilla sugar. Oil is melted in a porcelain mug, pre-soaked raisins and chopped nuts are placed in it. The mixture is heated for about 5 minutes over a moderate heat, stirring constantly. Pasta is boiled in milk, and vanilla and honey are added. After this, butter-nut sauce with raisins is poured on it. Served as a hot dish.

Noodles with honey

75 g honey, 80 g noodles, 15 g butter, 10 walnuts or almonds, salt. Boil the noodles in water. Then put them in a saucepan with melted butter. Add the not very finely chopped nuts and boiled honey. Mix and serve.

Milk noodles with honey

750 g milk, 50 g noodles, 1 tablespoon honey, 1 tablespoon butter, salt to taste. The noodles are poured into boiling milk. Salt and honey are added, and boiled until tender. Before serving, the dish is seasoned with butter.

Cottage cheese with honey

450 g of cottage cheese, 3 tablespoons of honey. Thoroughly pounded cottage cheese is mixed with honey to get a homogeneous mass, and served as an independent dish.

Cottage cheese casserole

50 g honey, 500 g cottage cheese, 1/4 cup sugar, 2 eggs, 3-4 tablespoons semolina, 1 tablespoon butter. Put honey, granulated sugar, eggs, semolina into the grated cottage cheese and mix thoroughly. This is then spread in a frying pan or in a dish greased with butter and baked in the oven for 35-40 minutes.

Curds with apples and honey (one portion)

20 g honey, 75 g cottage cheese, 30 g apples, 20 g sour cream, 10 g butter, 10 g sugar. Cottage cheese is passed through a meat grinder or rubbed through a sieve. The softened butter, grated peeled apples, sugar, and sour cream are combined with cottage cheese and mixed thoroughly. Before serving, it is recommended you pour slightly cooled butter over the curd mass with apples.

Cottage cheese casserole with honey (one serving)

30 g honey, 100 g cottage cheese, 15 g semolina, 60 g milk, 15 g sugar, 1 egg, 5 g butter, 5 g lemon zest, vanillin. Cook semolina porridge with milk to a medium thickness. After cooling, add an egg, sugar, salt, vanillin, lemon zest to it, stir, put in cottage cheese and mix thoroughly again. The prepared mass is transferred to a frying pan or in a mold, greased with an egg on top, sprinkled with powdered sugar, and baked in the oven. When serving, pour heated honey over it.

Curd mass with honey (one portion)

100 g cottage cheese, 10 g honey, 1 yolk, 15 g butter, 30 g sour cream or cream. Grind the yolk with granulated sugar and warmed honey, add softened butter, and beat until a homogeneous fluffy mass is formed. Then put grated cottage cheese into it, mix thoroughly again, and serve with sour cream or whipped cream.

Foods with honey

Rice with honey

150 g of honey, 200 g of rice, 50 g of raisins, 2-3 tablespoons of candied fruits, 50 g of walnuts, 150 g of cream, 1 tablespoon of powdered sugar. Rice is boiled in lightly salted water until tender, put in a colander, washed, and cooled. The sorted and washed raisins are mixed with sugar and put over a heat for a short time. Stir them while this is happening until the raisins swell slightly.

Then they're cooled. Raisins, honey, and half of the chopped nuts are added to the rice, and they are placed in portioned dishes - a salad bowl or a vase. When serving, the dish is sprinkled with the remaining candied fruits and nuts. Alternatively, it can be garnished with whipped cream and sugar.

Rice milk porridge with honey (one portion)

50 g rice, 70 g water, 50 g milk, 20 g honey, 10 g butter, 1 g salt. Before serving, melted butter is added to the porridge prepared in the traditional way, and honey is added.

Honey pudding

350 g honey, 400 g apples, 200 g corn flour, 200 g wheat flour, 40 g butter, zest and juice of 2 lemons, 1 tablespoon of baking soda, salt to taste. Add honey, wheat, cornflour, butter, soda, salt, and finely grated lemon zest to the apples cut into thin slices, and mix everything thoroughly. The prepared mixture is placed in a greased mold and baked in the oven for 1 hour.

Rice pudding with honey and oranges

3/4 cup rice, 1/2 cup honey, 3 cups milk, 2 tablespoons raisins, 2 eggs, 1/2 teaspoon salt, 1 tablespoon orange peel. Rice is boiled in milk for 45 minutes in a small saucepan under a lid in a water bath (add raisins 15 minutes before being ready). Beat eggs, add honey, salt, orange zest, and mix. The mixture is poured into rice and cooked for another 5 minutes, removing the lid, stirring constantly. After the pudding has cooled, it can be removed from the saucepan and served with tea or milk.

Oatmeal honey breakfast

500 g oatmeal, 2 cups milk, 3 tablespoons honey, 1/4 teaspoon salt, 250 g raisins, 200 g dried peaches, 125 g prunes, 2 medium apples, 200 g hazelnuts, 2 tablespoons lemon juice. Oatmeal is

mixed with milk, then salted, and refrigerated overnight. Before serving, add all the other ingredients (apple and peaches pre-soaked in water, as well as prunes cut as small as possible) and mix. Pieces of fresh fruit are placed on top and honey is poured on the dish before serving.

Rye bread crumbs pudding with honey and nuts

60 g rye crackers, 30 g honey, 1 egg, 30 g sugar, 10 g butter, 15 nuts, 5 g icing sugar, 10 g wheat flour. Grind the yolk with sugar, gradually adding softened butter, then add ground rye crackers and flour, chopped nuts, and stir thoroughly - you should get a homogeneous mixture. The whites, whipped into a thick foam, are introduced into it. Gently mix again from the bottom up. The prepared mixture is poured into a greased mold and sprinkled with breadcrumbs; it's then baked in an oven at a temperature of 230-250 ° C for 30 minutes. The pudding is taken out of the mold, transferred to a dish or flat plate. Cut out a depression in the upper part, fill it with honey, cover with a cut "lid", and sprinkle with powdered sugar on top.

Steam cottage cheese pudding with nuts and honey (one portion)

80 g cottage cheese, 20 g honey, 10 g flour, 1 egg, 15 nuts, 10 g butter, 15 g raspberries or strawberries, 10 g sugar. Combine heated honey, dried in a pan or in the oven, and chopped nuts, yolk, flour, butter, grated cottage cheese, and mix everything thoroughly. Introduce whipped protein and spread the mixture in a metal mold, greased with butter and sprinkled with granulated sugar. The mold is placed in a wide bowl with hot water (it should reach half the height of the container). Place a lid on top and heat the pudding at a low, steady boil for 25-30 minutes. The finished pudding is taken out of the mold hot, put on a plate, and

sauce is poured over it - this can be raspberries (strawberries) ground with sugar.

Pancakes with honey or sugar

250 g flour, 2.5 cups milk, 3 eggs, 2 tablespoons butter, 1/4 teaspoon salt, 2 tablespoons ghee for greasing the pan, 1 tbsp of sugar. Beat the egg yolks lightly, add 1/2 cup of milk, add salt, and stirring, add flour. Pour in melted butter, knead the dough so that there are no lumps, dilute it with the remaining milk (it is poured in gradually), and, at the end, add the egg whites whipped into a strong foam. Thin pancakes are cooked in a heated frying pan, greasing it with oil each time. Prepared pancakes are folded in four on a preheated dish and covered with a napkin. They are served with honey or sugar.

700 g flour, 500 ml baked milk, 15 g dry yeast, 1-2 tablespoons of honey, 1 tablespoon butter, 2-3 eggs, salt. Baked milk, flour, and dry yeast are stirred until the consistency of sour cream. When the dough rises, beat it with a mixer, add salt, honey, butter, eggs (whites need to be beaten into a strong foam) and let it rise a second time. After that, the dough is spread using a tablespoon in a greased frying pan and fried on both sides until golden brown. The pancakes are served with honey in the middle.

Glaze

Orange (peach) honey glaze

200 g light honey, 2 teaspoons orange (peach) syrup, 2 egg whites. The fruit syrup is cooled. The honey is heated in a water bath. The syrup is whipped with a mixer, honey is introduced into it in parts, and then the egg whites. This glaze is used to decorate honey cakes, gingerbread cookies, and small biscuits.

Chocolate glaze

100 g chocolate, 2 tablespoons honey, 100 g butter or margarine. Chocolate, butter (margarine), and honey are ground in a porcelain or glass bowl until smooth and this mixture is applied to the gingerbread.

Milk glaze

100 g honey, 1/2 l milk, 250 g butter, 100 g icing sugar, 2 tablespoons flour. Milk is mixed with flour, put in a water bath, a thick cream is boiled and set to freeze. The butter is ground with sugar and honey and gradually added to the chilled cream, after which it is thoroughly mixed.

Jams, desserts, and marinades

Honey jam

4 kg of unripe fruits or berries, 1 kg of honey. Honey is dissolved over low heat, boiled, berries or fruits are added. Honey jam is boiled in the same way as jam in sugar syrup. Sour berries (cranberries, cherry plums, etc.) require 1.5 times more honey than sugar. Honey jam sometimes turns sour, so it is not advisable to store it for a long time. You can cook jam with pure honey, and also with syrup consisting of honey and sugar taken in equal proportions.

Quince jam

1 kg of quince, 2 kg of honey. The quince is peeled, cut in half, cored, cut into slices, placed in a saucepan, topped up with a little cold water (the slices should be covered with water), and boiled until the quince is soft. Then the slices are transferred to a bowl. For cooking jam, honey and 1.5 cups of broth obtained by cooking quince are added to them. The jam is boiled over low heat until the quince slices become transparent.

Lingonberry jam

1 kg of lingonberries, 1 kg of honey, 3 pieces of cloves, a little cinnamon. Lingonberries are sorted out, washed, put through a sieve or colander, and allowed to drain. The berries are placed in a bowl for making jam. Honey is poured on. Cinnamon and cloves are added and boiled until tender.

Cranberry-apple-nut jam

1 kg of cranberries, 1 kg of apples, 1 glass of walnuts, 3 kg of honey, 1/2 glass of water. The cranberries are sorted out, washed, placed in a saucepan, and water is added. The mixture is boiled under a lid until the berries are soft. After that, they are rubbed through a sieve. Dissolve honey over low heat, let it boil, put mashed cranberries, peeled apples, nuts cut into slices, and boil for 1 hour.

Blackcurrant jam

1 kg of black currant, 2 kg of honey, 1 glass of water. Currant berries are sorted out, washed, put through a sieve, and allowed to drain. Water is added to the honey, brought to a boil, currants are added, and the jam is boiled over a low heat for 45 minutes.

Pickled grapes

1 kg of grapes, 50 g of sugar, 50 g of honey, 200 ml of vinegar, 200 ml of water, 20 g of salt, 5 cloves, 5 grains of cardamom. Rinse medium-sized ripe grapes, put them in a jar, and pour over them the marinade prepared from the indicated ingredients. Likewise, you can pickle plums, apricots, and other fruits and berries. Before pickling, fresh berries are often blanched (immersed in boiling water for a few seconds), and then cooled by immersing in cold boiled water for a minute.

Honey rings

125 g honey, 100 g sugar, 250 g flour, 1 egg, 1 teaspoon of spices, 100 g dried fruit. The honey is ground with sugar and egg. Flour is mixed with spices, and then with a honey-sugar-egg mixture.

Knead soft dough, roll it into a layer 1 cm thick, and cut out with two glasses of different diameters of rings. They are baked at a temperature of 175 ° C for about 15 minutes, after which they are glazed and decorated with finely chopped dried fruits.

Travel cookies

300 g flour, 150 g honey, 120 g sugar, 1 egg, 1/2 teaspoon of spices (cloves, cinnamon), 1/4 teaspoon of baking soda. All ingredients except honey, are mixed. The honey is heated and, after it has cooled, is introduced into the dough. It is rolled out into a layer 0.5 cm thick and cut out with cookie cutters. It is baked at a high temperature.

450 g of rye flour, 6 tablespoons of honey, 4 eggs, 5 g of ground cinnamon, 2-3 pieces of powdered cloves, a bag of vanilla sugar. All ingredients are mixed, the dough is rolled out (it should not be very thick) into a layer 0.5 cm thick, and circles are cut out with a thin glass. Cookies are baked in a moderately heated oven.

120 g honey, 100 g butter, 360 g flour, 70 g sugar, 2 eggs, 1 teaspoon soda, 1 tablespoon sour cream, almonds. The honey is heated and butter is melted in it. The mixture is removed from the heat, allowed to cool slightly, and flour, sugar, eggs, and soda are added to it. Knead the dough and leave it at room temperature for 1 hour. After that, it is rolled into a layer 1 cm thick, cookies are cut out with cookie cutters, smeared with egg yolk mixed with sour cream, and an almond kernel is placed in each cookie. Bake it in a moderately heated oven.

125 g honey, 1 kg oatmeal, 1/2 cup vegetable oil, 200 g wheat germ, 250 g bran, 125 g sesame seeds, 125 g sunflower seeds, 125 g coconut flakes, 1 tablespoon vanilla sugar. Oatmeal, bran, wheat germ, coconut, sesame and sunflower seeds are mixed. The honey is slightly heated. Butter, vanilla sugar are added, and this mixture is poured into the dry ingredients. Knead the dough, roll it into a 1 cm thick layer, place it in an oven preheated to 190

° C, bake for about 20-25 minutes, turn it over and dry it for another 2-3 minutes. The cooled layer is cut into small squares and sprinkled with powdered sugar.

Glazed honey cookies

250 g honey, 250 g rye flour, 300 g wheat flour, 100 g sugar, 4 eggs, 1 bag of baking powder. For icing: 2 egg whites, 200 g sugar, 5 teaspoons of water, 2 tablespoons of powdered sugar, a few drops of vinegar, salt on the tip of a knife. Rye flour and hot honey are stirred with a wooden spatula. Wheat flour, baking powder, sugar, and eggs are kneaded separately, and then added to the mixture of rye flour and honey. The dough (it should be of medium consistency) is left for a day. The following day, it is cut into 3-4 pieces, rolled into a layer 0.5 cm thick, and small round cookies are cut out in a stack or with special molds. Bake it at high temperature for 5-7 minutes. The finished cookies are put in a deep bowl or saucepan. Pour a warm glaze on them, slightly stirred so that they're evenly covered, and place in the oven with the door open - the products should dry out.

To prepare the icing, syrup is boiled from sugar and water, then egg whites whipped into a strong foam, vinegar, salt, and, last of all, powdered sugar is gradually added to it.

Honey cakes

50 g honey, 100 g sugar, 250 g flour, 4 eggs, 1 bag of baking powder, 1/2 teaspoon of cinnamon. Grind eggs with sugar, add warmed honey, cinnamon, baking powder, flour, and knead a thin dough. It is poured onto a greased, floured baking sheet, leveled with a wide-bladed knife, and baked in a moderately heated oven. The cooled product is cut into square or rectangular cakes.

250 g of honey, 400 g of flour, 100 ml of water, 200 ml of sunflower oil, 100 g of walnuts, 100 g of raisins, 2-3 teaspoons of candied fruits, 2 teaspoons of cocoa powder, 1/2 teaspoon of

cinnamon, 1 teaspoon of baking soda. Mix all the ingredients, knead the dough, roll it out in a layer of 2 cm and bake on a greased baking sheet in a moderately hot oven for 1 hour. Cover the cooled cake with chocolate icing or sprinkle with icing sugar.

400 g of honey, 500 g of flour (rye or wheat), 200 g of sugar, 5 eggs, 250 g of sour cream, spices (cinnamon, cloves, vanillin) to taste, 1 teaspoon of baking soda. Sugar, eggs, sour cream, spices, soda, and flour are added to the melted honey. The dough is poured into a greased mold and the cake is baked in a moderately heated oven.

Sour cream cake

250 g honey, 250 g sugar, 400 g flour, 3 eggs, 50 ml vegetable oil, 50 g walnuts, 1 teaspoon of baking soda. For cream: 700-800 g sour cream or cream, powdered sugar to taste, 50 g walnuts. Mix honey, sugar, and eggs to white, add vegetable oil, soda, ground nuts, flour, and knead the dough. It is divided into 5-6 pieces; each piece is rolled into a thin rectangular layer and baked in a moderately heated oven. The cooled cakes are placed on top of each other, lubricated with cream, and cut evenly. Sprinkle the finished cake with crushed crumbs.

Honey roll with nuts

50 g honey, 3 eggs, 100 g semolina, 100 g walnuts, 50 g sugar, 1/2 lemon zest, 1 glass of jam. Eggs are beaten with sugar, honey, and zest. Add semolina and minced nuts. The resulting mixture is poured onto a baking sheet, covered with parchment, and baked in a moderately heated oven until golden brown. Carefully remove the parchment, allow the product to cool, grease it with jam, and roll it up.

A vegan honey-based recipe

Extras

Milk eggs, beaten with honey

100 g honey, 9 eggs, 150 ml milk, salt. The honey is warmed up. When it becomes liquid, add eggs, milk, salt, and beat well. The resulting mixture is poured into a greased pan and scrambled eggs are baked in the oven.

Scrambled eggs with honey (one serving)

20 g honey, 2 eggs, 5 g butter. Grease the pan with butter and pour the eggs onto it so that the yolks remain intact. Egg whites are poured over with heated honey. Fry the eggs for 1–2 minutes, after which they are cooked in the oven.

Omelet with honey and nuts (one serving)

15 g honey, 2 eggs, 20 g wheat bread, 10 g nuts, 15 g butter, 30 ml cream. The crumbs of stale white bread are rubbed through a sieve or colander and lightly fried in a pan with butter along with finely chopped walnuts; stir throughout. Add honey. Beat the eggs with cream (sour cream can be used), add a mixture of bread crumbs with honey and nuts to them, stir, and immediately pour the mixture into a hot frying pan greased with butter. The finished omelet can be served as a dessert dish, garnished with slices of apples, pears, peaches, or apricots cooked in sugar syrup, or canned fruits.

Omelet stuffed with honey and nuts (one serving)

15 g honey, 2 eggs, 10 ml milk, 30 g sour cream, 15 g biscuit, 15 g nuts (pistachios, almonds or walnuts), 10 g butter. Cut the biscuit into small cubes, add finely chopped nuts and honey diluted with hot milk. Gently mix everything so that the biscuit pieces retain their shape. The eggs are mixed with sour cream, a little salt is added, and poured into a greased frying pan. The omelet is cooked over medium heat, stirring, until it thickens. After that, minced meat is added to the middle - a biscuit with nuts and honey - and it is closed on both sides, lifting the edges with a knife and giving the omelet the shape of a pie. The dish is then cooked in the oven and served immediately.

Omelet with biscuit, nuts, and honey

50–75 g of honey, 8–10 eggs, 125–150 g of sour cream, 50–75 g of biscuit, 50 ml of milk, 25–30 g of nuts, 3-4 tablespoons of butter, 1 tablespoon of powdered sugar. The stale biscuit is cut into cubes, mixed with chopped walnuts and honey, previously diluted with hot milk. Beat eggs lightly, add sour cream to them, salt, and knead well. The resulting mixture is poured into a preheated frying pan, greased with oil. When this thickens, put a mixture of biscuit, nuts, and honey in the middle. The edges of the omelet are wrapped on both sides in the shape of a pie and cooked in the oven. When serving, sprinkle the omelet with icing sugar.

Honey meringues

100 g honey, 300 g flour, 150 g icing sugar, zest of 1 lemon, spices (cinnamon, cloves) to taste, salt on the tip of a knife. Add cinnamon, crushed cloves, zest, salt, and powdered sugar to the flour. Lastly, add melted honey to the dough (it should not be very thick), roll it into a layer 0.5 cm thick, and cut out small circles with a glass. Bake the meringues on a greased baking sheet in a moderately heated oven.

Creamy honey

2 cups milk, 3/4 cup honey, 1/4 teaspoon salt, 2 eggs, 1 cup cream. The beaten eggs are carefully added to the hot milk and everything is mixed well. The mixture is boiled for 2-3 minutes in a water bath, allowed to cool, and lightly whipped cream is poured in. Stir again and refrigerate or ice.

Chocolate cream

150 g honey, 250 g butter, 100 g chocolate, cinnamon. The honey and chocolate are heated in a water bath - they must be thoroughly mixed. Remove from heat and continue stirring until the mixture has completely cooled. Add butter and cinnamon, stir again thoroughly, and beat the cream to the desired consistency (by the way, it is very good for finishing and decorating cakes).

Curd cream with carrots and honey

100 g honey, 200 g cottage cheese, 60 g carrots, 1/2 cup milk, 120 g apples, 1/2 orange, 1/2 tbsp. l. vanilla sugar powder, a little lemon zest. Milk, grated carrots and peeled grated apples are ground down in a mixer. Add cottage cheese, lemon zest, honey, vanilla sugar, beat again and chill in the refrigerator or on ice. When served, this cream is garnished with orange slices.

. . .

Strawberry cream with honey

100 g honey, 200 g butter, 1 yolk, 100 g icing sugar, 2 tablespoons of thick strawberry syrup. Honey is mixed with strawberry syrup. Butter is ground with icing sugar and yolk, gradually adding honey and syrup. Then the mixture is whipped into the desired consistency. This cream is good for adding to cakes, perhaps in the middle.

Curd-honey cream with raisins

100 g honey, 300 g cottage cheese, 30 g butter, 1 yolk, 1/2 tbsp. l. vanilla sugar powder, 10 g raisins, 100 ml milk, 200 g fruit syrup. Beat honey, butter, cottage cheese, milk, yolk, syrup, and vanilla sugar until the desired consistency is achieved, and gently mix the raisins into the cream.

Ice cream

750 g honey, 1 l cream. The cream and honey are mixed well and put in the freezer for a short time.

TWELVE

Improving the Honey Base

In areas of intensive farming, where many bee colonies are concentrated, the honey base for these insects becomes insufficient. To increase the honey reserves of the area, it is necessary to expand the availability of crops of complex use (for the needs of beekeeping, dairy cattle breeding, etc.). Excellent "honey plants" can be introduced in between food crops, or in collective and personal garden plots.

Close-up of a bee collecting pollen

Linden

Various types of linden (small-leaved, large-leaved, Manchurian, etc.) are most suitable for landscaping homesteads and farms, collective gardens, and other territories. The value of linden lies in the fact that its flowers emit a lot of nectar, they're used for medicinal purposes, and linden also creates a kind of aura favorable for human health. In addition, linden wood is used to make souvenirs.

Linden can be planted in autumn, winter, and spring. For the seeds to give good shoots, they are prepared for sowing using the following technique:

Initially, the seeds are soaked and kept in water at room temperature for 8 days. The water is changed after 2 days. Then the seeds are stratified. To do this, they are laid in washed river sand, sifted through a sieve, in a ratio of 1: 3 (1 part of seeds, 3 parts of sand), or mixed with sifted peat in the same proportion. The mixture is moistened to such a state that, when squeezed in a fist, droplets of water protrude from the peat mass, the sandy water is not squeezed out, and the lump of sand does not crumble.

The mixture is poured into wooden boxes measuring 50 x 30 x 15 cm with holes. The thickness of the soil layer should not exceed 15 cm. For 60–90 days, they are kept at a temperature of + 1–5 ° C, moistening and stirring the mixture every 8–15 days.

If the seeds have not sprouted 10-15 days before sowing, they must be transferred to a room with a temperature of + 25-30 ° C, gently stirring and moistening daily until they start growing.

Stratified seeds are sown in spring together with sand or peat. The planting depth of small-leaved linden seeds is 1.5–2 cm, large-leaved - 2–3 cm, and Manchurian - 3-4 cm. Seeds collected in September can be stratified for 30 days at a temperature of + 15–25 ° C and sown without signs of sprouting before winter. Autumn crops produce better shoots in spring. In winter, the seeds are very often destroyed by mice; therefore, spring sowing is more widely used. After sowing, the soil is mulched with peat or humus with a layer of 1 cm, as well as forest rotted leaves with a layer of 4–6 cm.

First, linden seedlings need shading. In cloudy weather, it is removed, and after the lignification of seedlings is stopped. In the first year, weeding is carried out with loosening to a depth of 3-5 cm, and in the second - to a depth of 10 cm (3-5 times). Row spacing should be at least 15–20 cm.

In 15–20 days after the emergence of seedlings, the plants are thinned out, removing the weak ones. On 1 m furrows leave 40-50 seedlings with a distance of 2-3 cm between them. In winter, they are covered with light ice to protect them from freezing.

Linden seedlings need a lot of moisture. Water them in the afternoon. If it's a dry autumn, it is imperative to carry out abundant watering.

At the age of three years, lindens are planted in a permanent place or transplanted for growing. The distance between the rows is 80 cm, and between the plants in the rows is 20–30 cm. Small-leaved linden seeds are harvested in September - October.

You can sow cucumber grass, snakehead, phacelia, mustard, and other strong honey plants.

Cucumber herb

This develops better when lime is added to the soil. Seeds are sown in spring or autumn in 2-3 terms. They are sown in a wide-row method with a row spacing of 45 cm. The seeding rate is 6 kg per 1 hectare. Seeds of cucumber grass germinate on the 5-6th day. The culture blooms in July - August. The flowers are blue, large, collected in a curl inflorescence. Honey productivity is 300-500 kg per hectare.

Snakehead

At the beginning of growth, the snakehead develops slowly, so it must be cultivated on clean soils (free of weeds). The crop is sown in early May in a wide-row method with a row spacing of 45 cm. The seeding rate is 5–6 kg per hectare. Blooms from July to late August. The flowers are blue-violet. Honey productivity - 200-400 kg per hectare.

If there are no large areas available, it is possible to generate "honey maker plants" in relatively small areas, in particular in gardens.

Sowing phacelia plants in the garden

Sowing phacelia between the rows of gardens is in the interests of both beekeepers and gardeners. This crop has a high reproduction rate, which distinguishes it favorably. It is also a good source of the classic green manure. Plowing the green mass of phacelia for fertilization leads to an increase in the yield of fruit plantations by 16-24%.

To obtain a higher yield of green mass, the seeding rate of these melliferous plants in orchards can be increased in comparison with the recommended one in this zone by 15–20%, that is, up to 15–16 kg per hectare. The plowing is done at the end of flowering (in August or September) after being used as a honey plant.

Lemon balm

Lemon balm grows well on light fertile soils. It can be propagated by seeds, seedlings, as well as cuttings. Seeds for seedlings are sown in boxes during the period of intense snow melting, covered with a layer of sand 4 mm thick, and covered with plastic wrap. To stratify the seeds, the boxes should be placed in melting snow during the first 3 days; make sure that they do not freeze. Then the boxes with seeds are transferred to an unheated greenhouse.

By the 10th day after sowing, the temperature in it is brought to +12 ° C; at the same time, the soil is slightly moistened (water temperature is +16 ° C). After emergence, the temperature is raised to + 18–20 ° C. The film is removed after the second pair of leaves appear. When the seedlings have four leaves, the seedlings need to be put into another box so that the plants are at a distance of 5–8 cm from each other. Seedlings are planted in a permanent place, keeping the distance between them in a row of 25-30 cm.

By the way, lemon balm is used as a medicine for many diseases.

Other honey plants

Sowing honey plants (phacelia, buckwheat, mustard) for honey harvest and green manure should be done in mid-June. Of course, the dates may vary depending on the weather conditions.

Melissa blooms in July - August. Her flowers are whitish-blue. Honey productivity is 150-200 kg per hectare.

Green hedges

Fences, which include tree and melliferous shrub plants, are important for protecting apiaries from the wind, and therefore for the development of beekeeping. Plantations with a predominance of such melliferous plants as linden, Norway maple, European chestnut, white acacia, walnut, and willow tree, create a stable honey base and enable bees to collect up to 1 ton of honey per hectare.

A lush hedge of white garden roses

To increase the profitability of apiaries, it's advisable to plant melliferous hedge plants. From tree and shrub species, you need to choose those that regularly bloom and bear fruit. Apricot, acacia, cherry plum, cherry, gleditsia, pear, honeysuckle (blue and Tatar), willow, Canadian irga, maple, buckthorn, almond, peach, Chinese apple tree bloom annually.

Actinidia, barberry, common honeysuckle, viburnum, dogwood, lemongrass, large-leaved linden, sea buckthorn, walnut, blackthorn, plum, cherry, mulberry, and Siberian apple trees also bloom, and, as such, are good alternatives.

Final Words

If you, dear reader, have decided to take up beekeeping, you will have an exciting journey into a world that most people know little about. You will gain knowledge about a fascinating topic and , step by step, year after year, you will move towards your goal.

You may be a little nervous starting out, but the time will come when that will pass. You will see that your movements have become precise and accurate. You will know what to do and when to do it. You won't need to rush to your books on beekeeping, frantically leafing through to find out what to do. You will be calm and joyful. And, even the sting of a bee brings you not pain, but health. In fact, if you do get stung, you might be more inclined to feel sorry for the bee that died in the defense of its home. They are strange little creatures, and their world and society, as with other insects, is alien in many ways to ours. But they are essentially harmless little beings and they bring their own blessing to the world. You'll soon come to regard them with affection, even with love, and you'll protect them as much as you can. When you see this in yourself, you'll know that you have become a true beekeeper.

Bees are all but invisible to millions of people who live in cities. But they're perceptible for those who live in the countryside. For those who decide to put even a few hives in their garden, life is firmly and forever connected with bees.

I sincerely hope that you decide to start your own hive. Bees may not realize it, but the fact is, they are the friends of mankind. So, I'm sure you will treat bees with the deep respect they deserve, as wonderful little beings who are here to help us.

Printed in Great Britain
by Amazon

77917201R00078